UNDERSTANDING

D.C. POWER SUPPLIES

Barry Davis

Royal Melbourne Institute of Technology

Chapter 1, 2.

Prentice-Hall of Australia Pty Ltd

A Prentice-Hall Direct Edition

National Library of Australia
Cataloguing-in-Publication Data

Davis, Barry, 1941-
Understanding D.C. power supplies.

For senior technical students.
Bibliography
Includes index.
ISBN 0 7248 1238 5.

1. Electric machinery - Direct current.
I. Title.

621.319'12

Prentice-Hall of Australia Pty Ltd, Sydney
Prentice-Hall International Inc, London
Prentice-Hall of Canada Ltd, Toronto
Prentice-Hall of India Private Ltd, New Delhi
Prentice-Hall of Japan Inc, Tokyo
Prentice-Hall of Southeast Asia Pte Ltd, Singapore
Whitehall Books Ltd, Wellington
Prentice-Hall Inc, Englewood Cliffs, New Jersey

Printed and bound in Australia by
Globe Press Pty Ltd

5 86

Contents

FOREWORD

PREFACE

1 RECTIFIERS *1*

1.1 Introduction *1*
1.2 Half-wave rectification *3*
1.3 Full-wave rectification *7*
1.4 Full-wave bridge rectifier *9*
1.5 Advantages and disadvantages of rectifier circuits *13*
1.6 Diode characteristics *13*
1.7 Faults in rectifier circuits *15*
1.8 Self-evaluation questions *18*

2 FILTERS *21*

2.1 Introduction *21*
2.2 Percentage ripple *22*
2.3 Half-wave rectifiers with capacitor filter *23*
2.4 Full-wave rectifiers with capacitor filter *25*
2.5 Choice of input capacitor *28*
2.6 L-C section filters *30*
2.7 CR section filter *33*
2.8 Voltage regulation *36*
2.9 Faults in filter circuits *37*
2.10 Self-evaluation questions *39*

3 VOLTAGE REGULATORS 41

3.1 Introduction 41
3.2 Zener diode regulator 42
3.3 Simple transistor shunt regulator 47
3.4 Series transistor regulators 48
3.5 Overload protection 55
3.6 Operational amplifiers in regulators 56
3.7 Current regulators 58
3.8 Self-evaluation questions 61

4 INTEGRATED VOLTAGE REGULATORS 63

4.1 Introduction 63
4.2 Packages 64
4.3 Connection diagrams and coding 66
4.4 Heatsinking 68
4.5 Basic three-terminal regulator 70
4.6 Applications 76
4.7 Dual power supply 83
4.8 Adjustable power regulators 89
4.9 Basic power supply circuits 93
4.10 Self-evaluation questions 97

5 VOLTAGE MULTIPLIERS 99

5.1 Introduction 99
5.2 Voltage doublers 101
5.3 Voltage tripler 104
5.4 Voltage quadrupler 108
5.5 Features of voltage multipliers 110
5.6 Self-evaluation questions 111

6 D.C. TO D.C. CONVERTERS 112

6.1 Introduction 112
6.2 Vibrator power supplies 113
6.3 Transistor converters 115

6.4 Reverse polarity protection *121*
6.5 Applications *121*
6.6 Self-evaluation questions *125*

7 SWITCHED-MODE POWER SUPPLIES *126*
7.1 Introduction *126*
7.2 Switching versus linear regulators *127*
7.3 Basic circuits *131*
7.4 Design information *140*
7.5 Control circuitry *145*
7.6 Protection circuits *153*
7.7 Three-terminal regulators in switched-mode circuits *161*
7.8 Self-evaluation questions *166*

8 FAULT FINDING AND TEST CIRCUITS *168*
8.1 Introduction *168*
8.2 Circuit measurements and component faults *169*
8.3 Safety precautions *171*
8.4 Component replacement and circuit board repair *173*
8.5 Fault-finding procedure and testing technique *175*
8.6 Fault diagnosis *182*
8.7 Diode protection *197*
8.8 Self-evaluation questions *199*

APPENDIX A Answers to self-evaluation questions *203*
APPENDIX B Glossary of terms and electronic symbols *214*
APPENDIX C Calculation of ripple and form factors for unfiltered half- and full-wave rectifiers *217*
APPENDIX D Calculation of average and R.M.S. values *221*

BIBLIOGRAPHY *227*

INDEX *229*

Foreword

The author's purpose in writing this text has been to provide an updating of material covering power supplies for the operation of electronic equipment. This area is a vital part of the training for students of electronics. This text is intended as a supplement to several of the excellent text books which cover the basic theory of electronics. It has a full coverage of the material required for the subject D.C. Power Supplies in Radio Tradesman's, Technician, and general electronic courses.

This book also deserves, and will certainly acquire, a wider audience from the many thousands of people who have a hobby interest in electronic projects and construction. All circuits shown have been tested by the Author, and would provide excellent practical projects for the student in the class-room and the hobbyist at home. These circuits include the very efficient, switched-mode system which is now being used in a great deal of new equipment coming on to the Australian market.

All those who are employed in the electronic industry have a responsibility to themselves, their customers, and their employers to become proficient in their trade. They have a duty to equip themselves with whatever knowledge is essential to safe and efficient performance, so they can carry out repairs and maintenance at reasonable cost. This text will add to that knowledge.

John R. Wales - M.A.C.E., S.A.I.R.E.E.
Head Telecommunications Division
R.M.I.T. Technical College
Melbourne

Preface

This textbook is intended primarily for radio/electronic apprentices, technicians, and students who already have a basic knowledge of semiconductor devices. It is a comprehensive study of power supplies, covering basic rectification, filtering, regulation and voltage multiplication. It then explores current 'state of the art' technology, with an in-depth look at integrated voltage regulators and switched-mode power-supply circuits.

The book should prove invaluable to the student who is preparing to take any examination which includes power-supply principles. Self-evaluation questions are included at the end of every chapter and solutions are given in an appendix. The questions are designed to test the level of understanding achieved from each chapter.

Every effort has been made to present the theory in a way that is logical and easy to follow. The use of mathematics has been kept to a minimum, so that the text will serve a wide range of readers. In addition the many practical circuits and design details will be of considerable interest to the hobbyist and home experimenter. The final chapter discusses fault-finding techniques and procedures.

Acknowledgements. The development of this publication was supported by funds from T.A.F.E. Particular Purpose Grants administered by the Royal Melbourne Institute of Technology (R.M.I.T.) Technical College, Curriculum Research and Development Committee, Melbourne, Australia. Without the help of this committee, this book would still be in its infancy.

Many people have contributed to the development of the book, and every effort has been made to acknowledge such assistance and stimulation in the bibliography.

Special thanks go to the technical reviewers Mr. D.L. Dare, T.Tr.I.C., Dip. Tech. Tchg., and Mr. P.L. Stark, T.Tr.I.C., Dip. Tech. Tchg., from the Telecommunications Division of R.M.I.T. Thanks also to Gloria Shanks for typing the manuscript, and to Jocelyn Smith for the graphics.

Finally, I should like to acknowledge my wife's patience and persistence in keeping my nose to the grindstone when I could easily have 'gone fishing'.

Barry Davis
Dip. Tech. Tchg. (S.C.V.)
T.Tr.I.C.
M.T.E.T.I.A.

1 Rectifiers

1.1 INTRODUCTION

Electric power is generally available in Australia, New Zealand, the United Kingdom, and South Africa in the form of an alternating supply at a fixed frequency of 50 hertz. The nominal voltage is generally 240 volts, and this represents its 'effective' (R.M.S.) value. However, a D.C. voltage is usually required to operate most pieces of electronic equipment; for example, a stereo amplifier may require 24 volts D.C. for satisfactory operation. Other types of equipment may require a higher or lower voltage than this, but the need for a *steady* voltage is common in all fields of electronics. D.C. power supplies are the circuits that change the 240 volts 50 Hz (A.C.) mains supply into the D.C. voltage we require.

There are four basic functions to be considered when discussing power supply circuits. They are:

1. *Voltage transformation:* the use of a transformer to step the 240 V 50 Hz input voltage up or down to the level required. It does *not* change the 240 V A.C. to D.C.

2. *Rectification:* the conversion of the A.C. voltage, with alternating polarity, into a voltage of constant polarity. The result is a pulsating D.C. voltage.

3. *Filtering:* the process of 'smoothing out' the fluctuations in the rectified voltage to give a 'smooth' D.C. voltage.

4. *Regulation:* the ability of the power supply to maintain a fixed level of D.C. voltage with a varying load.

This chapter discusses basic *rectifier-circuit* arrangements, their operation and expected output waveforms. The requirements, advantages, and disadvantages of the circuits are compared, and examples of relevant calculations are given. In Chapter 2, we consider how the pulsating D.C. voltage is applied to a *filter circuit* to smooth out the fluctuations. The types of circuits available to perform this function are shown, and their operation, merits and effect on circuit performance are also discussed. *Regulation* is a very important parameter in a power supply, and *voltage regulators*, both discrete and integrated, are given thorough treatment in Chapters 3 and 4. Basic regulator circuit operation is discussed in Chapter 3, and this leads to a consideration of three-terminal I.C. regulator circuits in Chapter 4. Handling precautions and circuit design ideas are also discussed then.

A list of notation used in this chapter is shown in Table 1.1.

Table 1.1 Abbreviations

V_p	= Voltage across primary (R.M.S.)
V_s	= Voltage across secondary (R.M.S.)
V_{pm}	= Maximum or peak primary voltage
V_{sm}	= Maximum or peak secondary voltage
V_{Lm}	= Maximum voltage across the load
V_L or V_{out}	= Average load voltage
I_{Lm}	= Maximum or peak load current
I_L	= Average load current
P.I.V.	= Peak inverse voltage (the reverse voltage rating of a diode)

A complete list of abbreviations and symbols used in the text and in the circuit diagrams can be found in Appendix B at the back of the book.

Note: the term 'current' in this text refers to *electron flow*, i.e.

from negative to positive, unless otherwise stated. This is in accordance with standards set down by the Electronic Trades Standing Committee.

1.2 HALF-WAVE RECTIFICATION

A PN junction, or semi-conductor diode, presents a low resistance to current in one direction and a high resistance to current in the opposite direction. Thus, it can perform rectification. Many types of semi-conductor diodes are available, and they vary is size from the very tiny, as used in sub-miniature circuitry, to very large, 500 ampere rectifiers, as used in large power supplies.

A simple half-wave rectifier circuit using a semi-conductor diode is shown in Figure 1.1. V_p is the generator voltage representing 240 V A.C. and is applied to the transformer primary. The transformer is connected so that no phase reversal takes place between primary and secondary (indicated by the dots at the top of the windings).

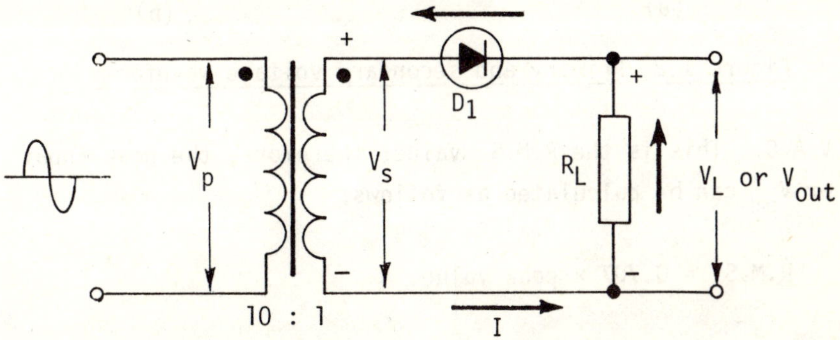

Figure 1.1 Half-wave rectifier circuit

When the top of the primary is made positive by the input signal, the induced voltage will make the top of the secondary positive. This makes the anode of D_1 positive, forward biasing it, and D_1 conducts. Electrons flow from the negative end of the transformer secondary winding up through R_L, through the diode and back to the top of the secondary winding. This current causes a voltage drop across the load resistor, with a negative polarity at the bottom and positive at the top. If we assume that D_1 is an 'ideal diode', (that is, it behaves as a short circuit when forward biased),

V_{out} will be equal to V_s. In practice, an ideal diode is not available, and therefore V_{out} or V_L will be less than V_s by the amount of voltage dropped over the diode (approximately 700 mV).

On the negative half cycle of input voltage, the top of the secondary winding is made negative. This reverse biases D_1, and the diode does not conduct. An output is available only when the input voltage forward biases the diode.

The transformer in Figure 1.1 is shown as a 10:1 step down. The waveforms for primary and secondary voltages for this are shown in Figure 1.2.

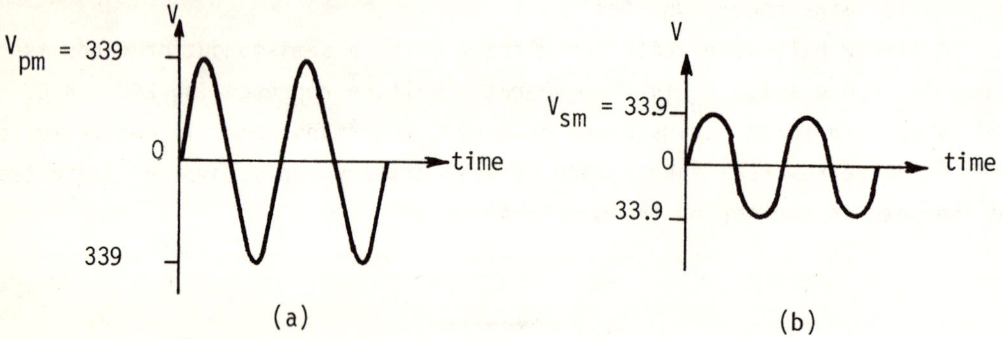

Figure 1.2 Primary and secondary voltage waveforms

V_p is 240 V A.C. This is the R.M.S. value; therefore, the peak input waveform, or V_{pm} can be calculated as follows:

$$\text{R.M.S.} = 0.707 \times \text{peak value}.$$

Therefore, the peak value equals R.M.S./0.707, or the R.M.S. × 1.414:

$$240 \times 1.414 = 339 \text{ volts}.$$

The transformer will step the peak voltage down to 1/10 of V_{pm}, making $V_{sm} = 339/10 = 33.9$ V.

The transformer has thus changed the magnitude of the A.C. voltage to the desired level, and this is applied to the diode. As explained above the diode conducts only on the positive half of the cycle, and the output waveform shown in Figure 1.3 results. Since only one half of the input waveform is represented in the output, this circuit is known as a 'half-wave rectifier'.

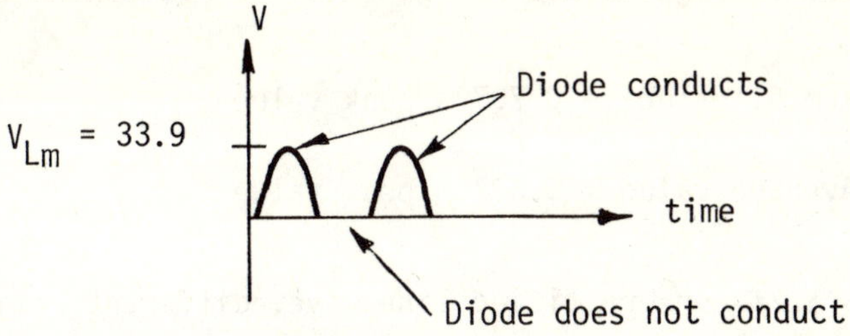

Figure 1.3 Output waveform of a half-wave rectifier circuit

Care must be taken in power supply circuits not to exceed the maximum current and voltage ratings of the components used. We need to consider the capabilities of the load, the diode, and the transformer. First, the current passing through the load will be considered. The peak load voltage in the example is 33.9 volts and if we assume the value of R_L to be 200 ohms, the peak load current can be found thus:

$$I_{Lm} = \frac{V_{Lm}}{R_L}$$

$$= \frac{33.9}{200}$$

$$= 0.1695 \text{ amperes} = 170 \text{ mA.}$$

Since the current through the load must follow the variations in V_L, it will also be half-wave. Figure 1.4 shows the output voltage and current waveforms with amplitudes.

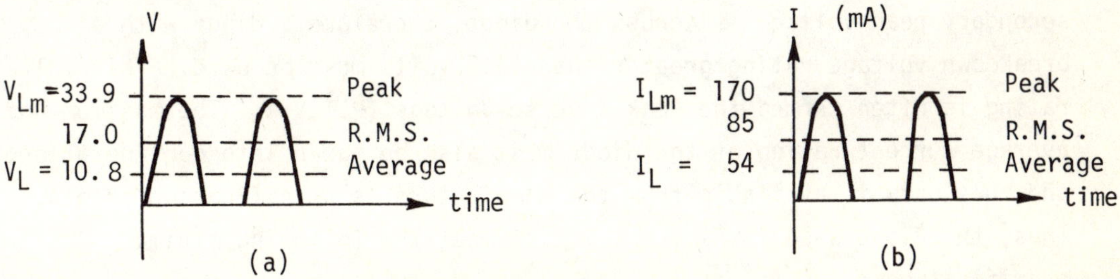

Figure 1.4 Half-wave output waveforms

For a *full* sine wave (current or voltage) the relationships between peak, R.M.S. and average values are given by the following formulas:

$$\text{R.M.S. value} = 0.707 \times \text{peak value}$$

$$\text{Average value} = 0.637 \times \text{peak value}$$

However, for a *half-wave* rectified sine wave, different relationships apply, because one half of the input waveform is clipped off. (For discussion and derivation of these and other formulas, see Appendix D.)

$$\text{R.M.S.} = 0.5 \times \text{peak value}$$

$$\text{Average} = 0.318 \times \text{peak value}$$

The power dissipated in the load can be calculated from the R.M.S. values of load voltage and current:

$$\begin{aligned}\text{Power} &= V \times I \\ &= 17 \times 85 \times 10^{-3} \\ &= 1.45 \text{ watts.}\end{aligned}$$

R_L should therefore have a power rating greater than 1.5 watts.

The diode must also be considered from the standpoint of its breakdown voltage. It conducts on one half-cycle only and is non-conducting (off) on the other half cycle. During this time the input voltage (V_S) reverse biases the diode, and there is no voltage across R_L. The full transformer secondary peak voltage is across the diode; therefore a diode with a breakdown voltage rating greater than 33.9 volts must be used. This voltage rating is often termed the Peak Inverse Voltage (P.I.V.). The maximum and average current rating of the diode must also be taken into consideration, and this data is available from the manufacturer's specification sheets. Thus, the diode used in Figure 1.1 must meet the following minimum specifications:

Peak current = 170 mA

Average current = 54 mA

P.I.V. = 33.9 volts

Finally, the transformer has a maximum performance rating also. This is usually specified in terms of a V.A. (volt-ampere) rating, which indicates the *product* of the R.M.S. voltage and current that can be safely handled. Calculation of the V.A. rating for the transformer used in Figure 1.1 is as follows:

$$V_{R.M.S.} \times I_{R.M.S.} = 17 \times 85 \times 10^{-3}$$

$$= 1.45 \text{ V.A.}$$

1.3 FULL-WAVE RECTIFICATION

A full-wave rectifier permits a current path in the same direction through the load for the full cycle of input. Two silicon diodes used in the circuit configuration shown in Figure 1.5 will perform full-wave

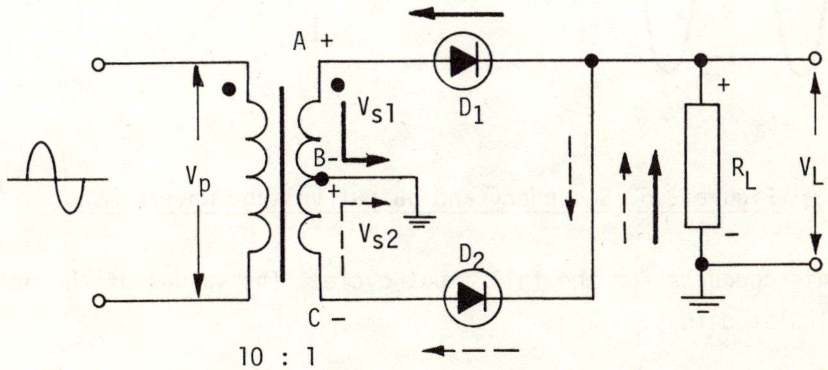

Figure 1.5 Full-wave rectifier circuit

rectification. An alternating voltage is fed to the primary of the transformer. When Point A on the transformer secondary is positive, Point C will be negative. D_1 will be forward biased, and D_2 will be reverse biased.

Electrons will flow from the centre-tap to earth, up through the load, through D_1, and back to the top of the transformer. This current path is indicated by the heavy arrows.

On the next half cycle, Point A will be driven negative and Point C positive; therefore D_1 will be reverse biased, and D_2 will conduct. Electrons flow from the centre-tap into earth, up through R_L, through D_2, and back to the bottom of the transformer. This current path is indicated by the dotted arrows.

It is important to note that the input voltage to each diode is one-half of the overall secondary voltage, and that the output waveform has the same polarity regardless of which diode is conducting. The input and output waveforms are shown in Figure 1.6. The primary voltage is 240 V R.M.S. and if a 10:1 step-down transformer is used again, the full V_s will be 33.9 volts. This is divided into two to provide each diode with its own secondary voltage. Therefore V_{s1} and V_{s2}, will each be 16.95 V. A full-

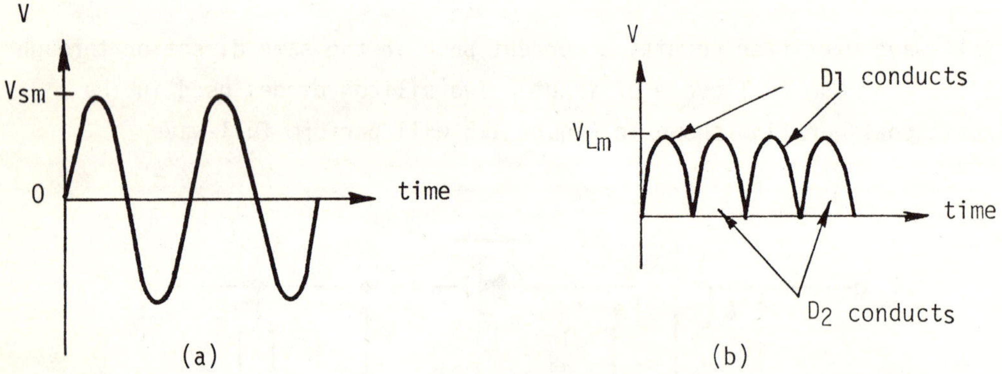

Figure 1.6 Secondary and output voltage waveforms

wave circuit conducts for the full input cycle. The values of the voltages can be calculated thus:

$$V_{R.M.S.} = 0.707 \times \text{peak value}$$

$$V_{average} = 0.637 \times \text{peak value}$$

The peak load current will be:

$$I_{Lm} = \frac{V_{Lm}}{RL}$$

If R_L is assumed to be 200 ohms, then:

$$I_{Lm} = \frac{17}{200}$$

$$= 0.085 \text{ amperes}$$

$$= 85 \text{ mA}$$

Figure 1.7 shows the output voltage and current waveforms with amplitudes.

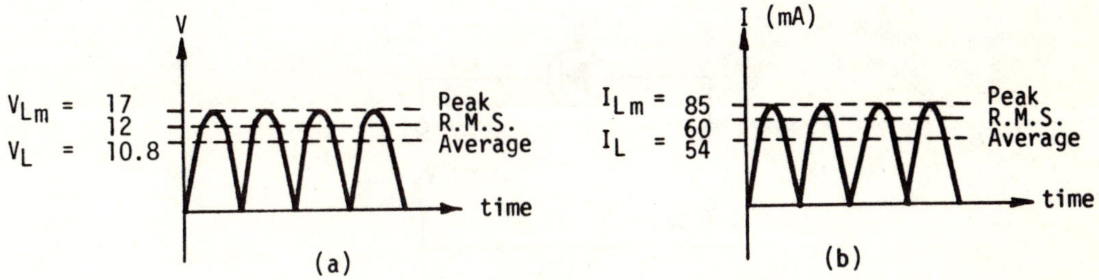

Figure 1.7 Full-wave output waveforms

To calculate the P.I.V. rating for a full-wave circuit, we must take into account the fact that one diode conducts while the other is off. For example, in Figure 1.8, when point A is positive, D_1 is conducting, D_2 is off, and $V_L = V_{s1}$. D_2 then has V_{s1} *in series with* V_{s2} across it (as shown in Figure 1.9) because D_1 is forward biased, and virtually no voltage is dropped across it. Therefore, each diode must have a reverse voltage rating (P.I.V.) equal to at least twice the peak load voltage.

1.4 FULL-WAVE BRIDGE RECTIFIER

A full-wave bridge rectifier uses a configuration of four diodes, in which two diodes conduct on one half of the cycle, and the other two conduct on

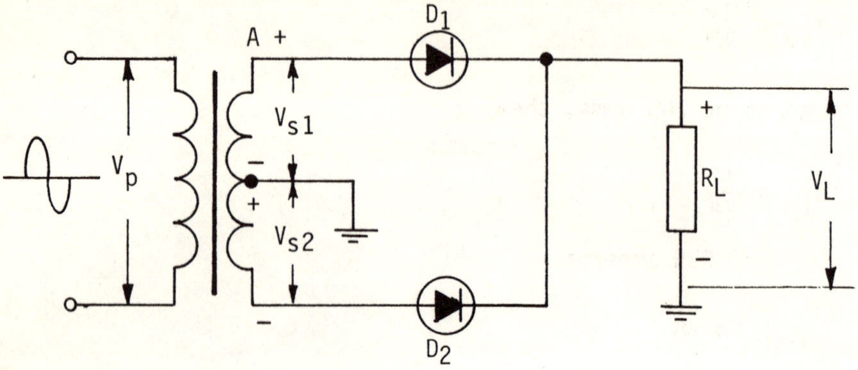

Figure 1.8 Full-wave rectifier

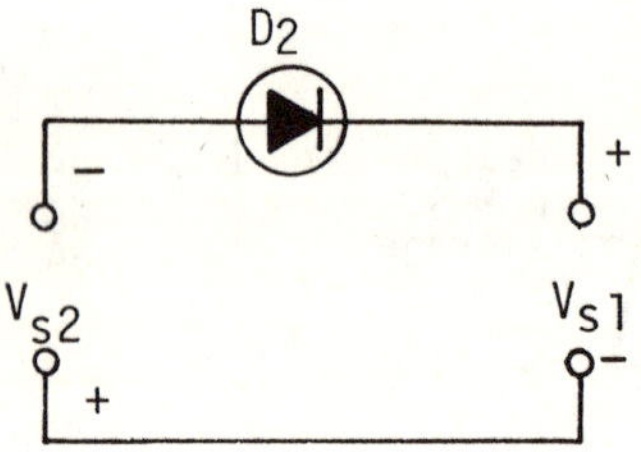

Figure 1.9 Voltage across the non-conducting diode

the next half of the cycle. An example is shown in Figure 1.10. The transformer secondary is connected across the bridge at Points A and B, and the output is taken across Points C and D.

When the top of the transformer is positive, Point A will be positive, and Point B will be negative. This means that the D_3 anode is positive, and the D_4 cathode is negative - i.e., forward biased. D_1 cathode is positive, and D_2 anode is negative, or reverse biased. Therefore, on the positive half cycle D_3 and D_4 conduct, and on the negative half cycle D_1 and D_2 conduct. Electrons flow from the bottom of the transformer through D_4 to the common line, up through R_L, and through D_3 to the top of the transformer. This path is indicated by the heavy arrows.

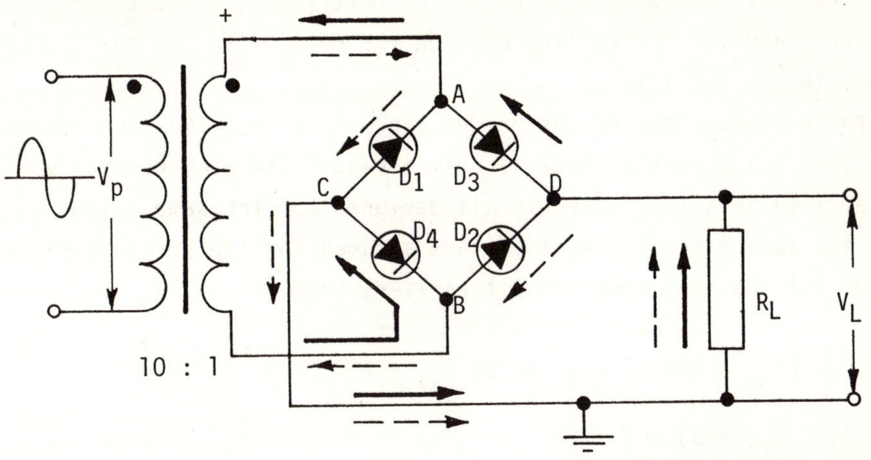

Figure 1.10 Full-wave bridge rectifier circuit

On the next half cycle, D_3 and D_4 are off, and electrons flow from the top of the transformer through D_1 to the common line, up through R_L, and through D_2 to the bottom of the transformer. This path is indicated by the dotted arrows. Current passes through the load in the same direction for both half cycles of input voltage. The input and output waveforms are shown in Figure 1.11.

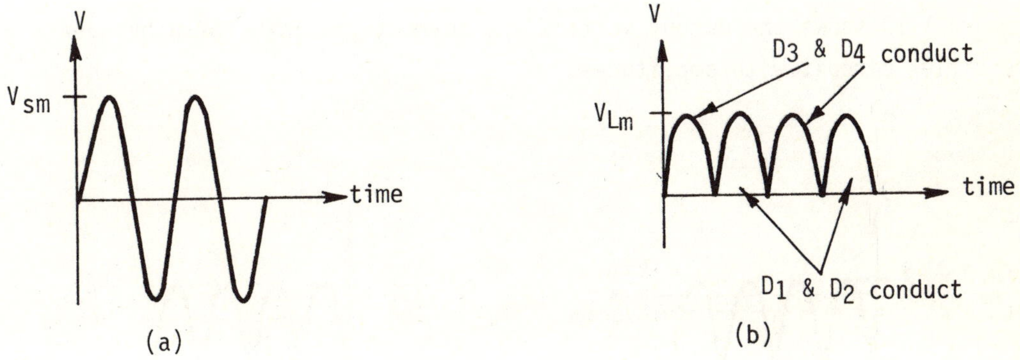

Figure 1.11 Secondary and output voltage waveforms

One advantage of the bridge rectifier is that, with a given transformer, the bridge circuit produces a voltage output nearly twice that of the full-wave circuit. This increase in voltage is due to the fact that the bridge does not use a centre-tapped secondary. The bridge circuit therefore applies the full secondary voltage to the rectifiers, while the full-wave circuit only applies the voltage between the centre-tap and one end of the secondary to the rectifiers.

A further advantage of the bridge circuit is that the peak reverse voltage (P.I.V.) across a diode is only one-half the peak reverse voltage across a diode in a full-wave circuit designed for the same output voltage.

If the same circuit conditions are assumed for the bridge circuit as were used for the full-wave circuit earlier, then:

$$V_{pm} = 240\ V_{R.M.S.} = 339\ V_{peak}$$

and

$$V_s = 33.9\ V$$

The load current is therefore:

$$I_{Lm} = \frac{V_{Lm}}{R}$$

$$= \frac{33.9}{200}$$

$$= 0.1695 \text{ amperes}$$

$$= 170 \text{ mA}$$

Figure 1.12 shows the output voltage and current waveforms of a bridge-rectifier circuit with amplitudes.

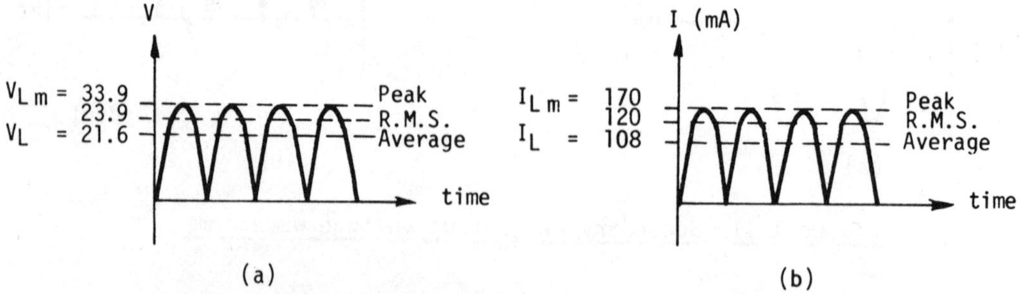

Figure 1.12 Full-wave bridge output waveforms

1.5 ADVANTAGES AND DISADVANTAGES OF RECTIFIER CIRCUITS

If we compare the three rectifier circuits given in this chapter under the same transformer voltage and circuit load conditions, then it can be seen that the full-wave bridge circuit gives the highest average output voltage and current levels.

While the half-wave circuit has the advantage of simplicity and low cost, it also has several disadvantages. It is not very efficient since only half the input wave is used, and the average output voltage is low.

The full-wave circuit is more efficient than a half-wave circuit because it operates on both half cycles of the input wave. Its disadvantage is that it requires a centre-tapped transformer, and the P.I.V. of the diode must therefore be higher.

Finally, the bridge rectifier, even though it requires four diodes, is relatively inexpensive. The transformer does not require a centre-tap, and given the same secondary input voltage, the output voltage will be twice that of a full-wave circuit. However, the bridge circuit is less efficient than a centre-tap circuit, particularly at low voltages with high load currents.

1.6 DIODE CHARACTERISTICS

Diode characteristics have their own language of abbreviation. Unfortunately, there are no standardized abbreviations used by all manufacturers. Table 1.2 shows diode terminology and the most commonly used abbreviations.

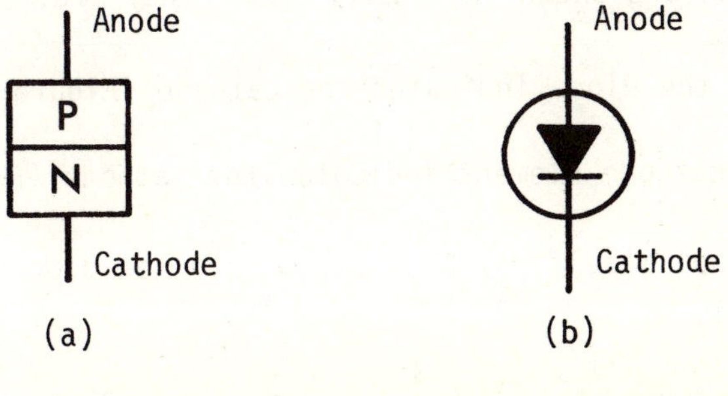

Figure 1.13 Diode structure and symbol

Table 1.2 Diode Terminology

Average forward current	$I_{F(AV)}$	The current to be handled by the diode under normal operating conditions
Peak forward current	I_{FM}	The absolute maximum current at any instant (often called I_{peak}).
Surge current	I_{FSM}	The maximum current the diode can handle for one second.
Forward voltage	V_{FM}	The operating voltage at a particular forward current. ($I_{F(AV)}$)(often called V_{peak}).
Reverse current	I_{RM}	The maximum reverse current at maximum reverse voltage.
Power dissipation		Permissible power at 25°C; the power rating is less above 25°C.
Peak reverse voltage		(P.I.V.) - the maximum voltage the diode can stand when in its non-conducting state, also referred to as the maximum peak reverse voltage, V_{RM}.

1.6.1 IDENTIFICATION OF THE CATHODE

The cathode of a diode is N-type semiconductor material, and the anode is P-type material. The relationship between this physical structure and the commonly used symbol is illustrated in Figure 1.13a. The symbol for the diode is derived for its semi-conductor structure (see Figure 1.13b). The bar represents the cathode,and the arrow represents the anode. *Electron flow is from cathode to anode.*

Several methods are used to identify which end of the diode is the cathode. Three of these are shown in Figure 1.14. They are:

1. A band around the diode indicates the cathode (Figure 1.14a).
2. A dot at the cathode element indicates the cathode (Figure 1.14b).
3. The diode symbol itself may be stamped onto the diode casing (Figure 1.14c).

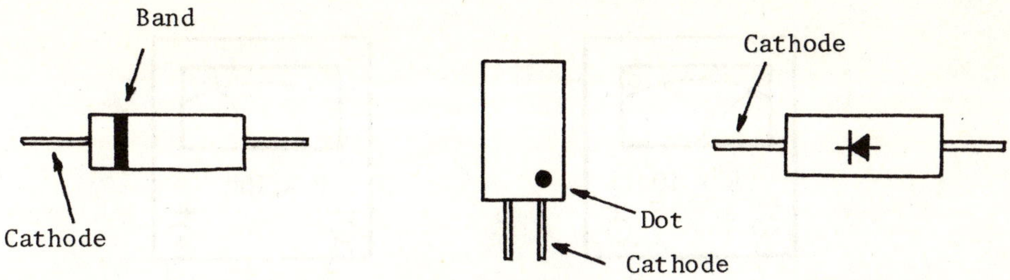

(a) Band (b) Dot (c) Symbol

Figure 1.14 Cathode identification

1.7 FAULTS IN RECTIFIER CIRCUITS

We will now consider possible faults in the components common to the power-supply rectifier circuits discussed in this chapter: diodes and transformers.

1.7.1 DIODES

Diodes are semi-conductor devices and are therefore susceptible to current surges. The faults encountered with diodes are of three types:

1. Open circuit
2. Short circuit
3. High resistance

All of the conditions can be caused by excessive current. An open-circuit condition exists when the PN junction of the diode has been blown apart. A back-to-front ratio check with an ohmmeter reads infinity in both directions. When the PN junction of the diode is fused together, it creates a short-circuit condition. A back-to-front ratio check gives a reading of zero ohms in both directions.

A condition of high resistance can exist within the diode if the junction has been partially damaged. The diode will then have a higher than normal voltage drop in the forward direction due to the higher level of resistance, $V = I \times R\uparrow$. The back-to-front ratio will reveal this by giving a reading of a lower than expected output voltage.

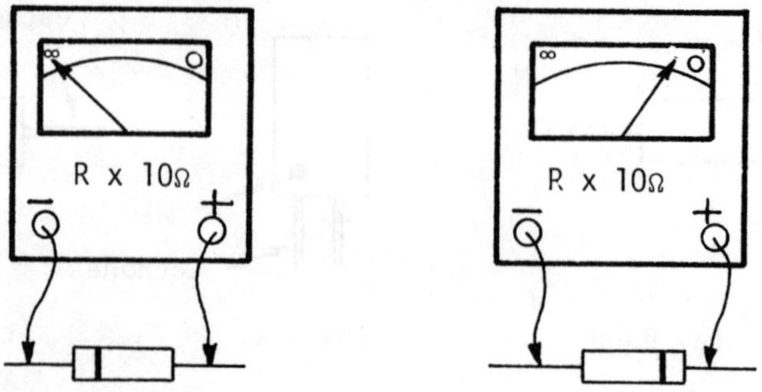

(a) reverse direction (b) forward direction

Figure 1.15 Circuit connection for checking the back-to-front ratio of a diode

Figure 1.15 shows the multimeter connections for checking a diode. *Most multimeters have a negative polarity at the positive socket.* This being the case, in order to forward bias the diode, the cathode end of the diode (indicated by the bar) must be connected to the negative polarity.

In the forward direction (cathode negative and anode positive) and with the ohmmeter range set to R × 10, a forward resistance of approximately 200 ohms will be shown. With the diode reversed, the reading will be infinity, ∞. The readings in the forward direction vary between diode types, i.e. germanium, silicon, signal or power diodes. However, the ratio of high reverse resistance and low forward resistance is true for all.

The forward resistance of the diode is inversely proportional to the bias voltage across it. This being the case, when a diode is operational in a power-supply rectifier unit, its forward resistance is much less than the tested resistance of 200 ohms. The voltage drop across the forward biased diode is therefore minimal.

1.7.2 TRANSFORMERS

Transformers are basically two coils of wire with magnetic linkage between them. Faults experienced are:

1. Open circuit primary winding, where an ohmmeter check reads infinity ohms.

2. Open circuit secondary winding, where again the ohmmeter reads infinity.

3. Shorted turns in either the primary or secondary winding.

The D.C. resistance of the transformer primary and secondary is quite low, therefore, ohmmeter checks must be carried out on the R × 1 range.

If a shorted turn exists, the D.C. resistance of the winding will be changed from its normal value. However, more importantly, the inductance of the coil will be reduced. The inductance is reduced because the field in the primary winding of the transformer is partially cancelled by the current induced into the shorted turn. The transformer operates with an A.C. voltage. With a reduced inductance the A.C. current in the coil will be higher, and this will result in the transformer getting physically hot.

1.8 SELF-EVALUATION QUESTIONS (ANSWERS IN APPENDIX A)

1. Figure 1.1 shows the polarity of the output voltage across the load resistor R_L. How can this polarity be reversed?

2. A rectifier circuit is named according to the way it performs.

   (a) A half-wave rectifier passes ..... of the A.C. input.

   (b) A full-wave rectifier passes ..... of the A.C. input.

3. What is the anticipated forward voltage drop across a diode?

4. For the circuit in Figure 1.1, assume V_{pm} = 163 V, the transformer turns ratio, primary to secondary is 1.15 : 1, and the load is 10Ω. Determine:

   (a) The R.M.S. secondary voltage of the transformer.

   (b) The minimum reverse voltage rating of the diode.

   (c) The peak and R.M.S. load current.

   (d) The power dissipated by the load.

5. What is the 'average' value of the output from a half-wave rectifier if the peak-to-peak secondary voltage is 100 V?

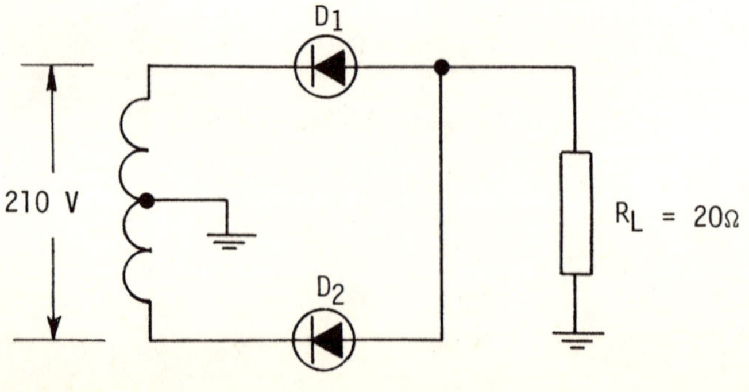

Figure 1.16

6. (a) What is the polarity of the output voltage in the circuit shown in Figure 1.16?

(b) Calculate the following:

(i) Peak secondary voltage for D_1 and D_2.

(ii) The average output voltage.

(iii) The average load current.

(iv) P.I.V.

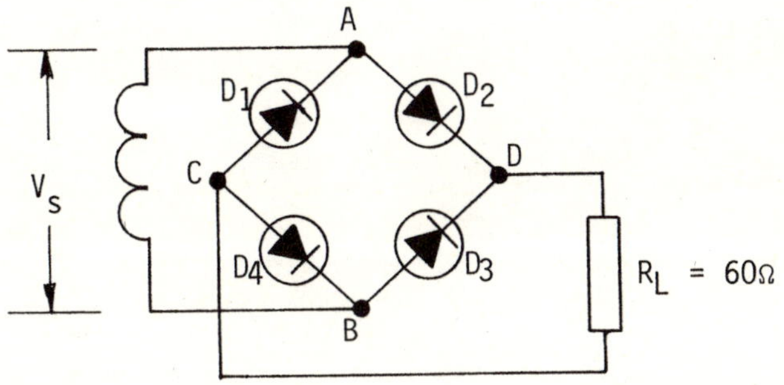

Figure 1.17

7. (a) What type of circuit is shown in Figure 1.17?

(b) Which diode or diodes conduct when Point A is positive with respect to Point B, and what is the polarity of the voltage at Point D with respect to Point C?

(c) Assume V_s is 40 volts R.M.S.; calculate the value of:

(i) The average output voltage.

(ii) The current through the load.

(iii) The peak inverse voltage rating of the diodes.

8. Assume the secondary voltage rating of a transformer was 250 V - 0 - 250 V, and it was used in a full-wave circuit configuration.

   (a) Calculate the peak voltage across the load and the P.I.V. of the diodes.

   (b) If the circuit was changed to a bridge configuration, would the values found for Part (a) remain the same? Explain.

9. Briefly explain the fault conditions you could experience with:

   (a) Diodes

   (b) Transformers.

2 Filters

2.1 INTRODUCTION

As shown in Chapter 1, alternating current can be converted into a pulsating direct current by the use of rectifier circuits. However, most electronic equipment requires a 'smooth' D.C. supply, and this can be accomplished by the use of properly designed filters.

The unfiltered output of a full-wave rectifier is shown in Figure 2.1.

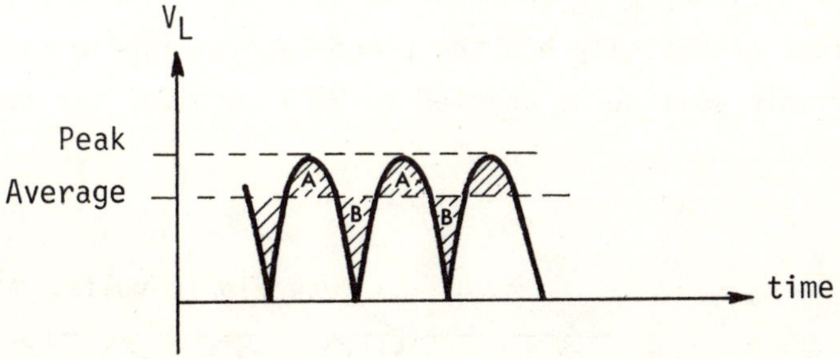

Figure 2.1 Unfiltered output of a full-wave rectifier

The polarity of the output voltage does not reverse, but its magnitude fluctuates about an average level as successive pulses of energy are delivered to the load. The average voltage for the waveform shown in Figure 2.1 can be defined as the level at which Area A is equal to Area B. In this case the average level is 0.637 of the peak value. The voltage fluctuations above and below this average value are called 'ripple'. The frequency of the ripple for a full-wave circuit is twice the frequency of the voltage that is being rectified. In the case of 50 Hz mains, the ripple is 100 Hz. For a half-wave circuit the ripple has the same frequency as the input voltage, i.e. for 50 Hz mains the ripple is 50 Hz. The higher the ripple frequency, the easier it is to filter the power supply; the fluctuations occur at a faster rate, and the time period between pulses is therefore reduced.

2.2 PERCENTAGE RIPPLE

The output of a rectifier is composed of (1) direct voltage and (2) an alternating (or ripple) voltage. The amount of ripple that can be tolerated varies with different applications and is designated by a percentage (see Appendix C).

The 'percentage of ripple' is 100 times the ratio of the R.M.S. value of the ripple voltage at the rectifier output to the average value of the total output voltage, V_{out}. That is

$$\% \text{ ripple} = \frac{V_{RMS}}{V_{out}} \times \frac{100}{1}$$

where $V_{RMS} = 0.707\ V_r$, and V_r is the peak value of the ripple voltage. Figure 2.2 indicates graphically how the percentage of ripple may be determined (the ripple voltage is assumed to be sinusoidal for purpose of explanation).

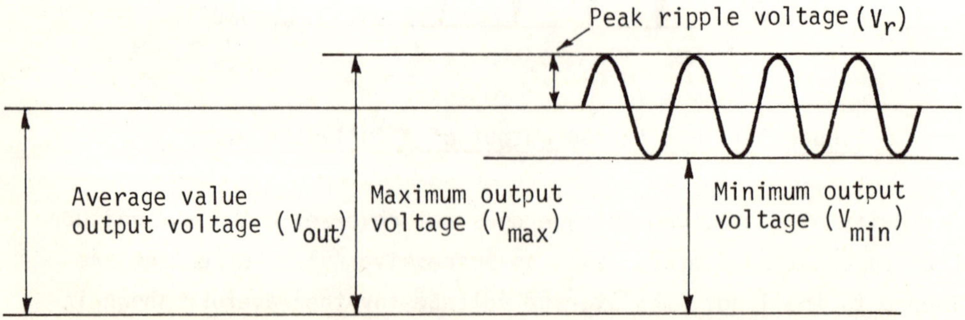

Figure 2.2 Output voltage levels with ripple

2.3 HALF-WAVE RECTIFIER WITH CAPACITOR FILTER

To obtain a reasonably smooth voltage output we require an electronic component that will oppose changes in voltage. A capacitor is such a component, and it becomes our basic filter. Figure 2.3 shows a half-wave rectifier circuit with a capacitor across the output. Adding the capacitor

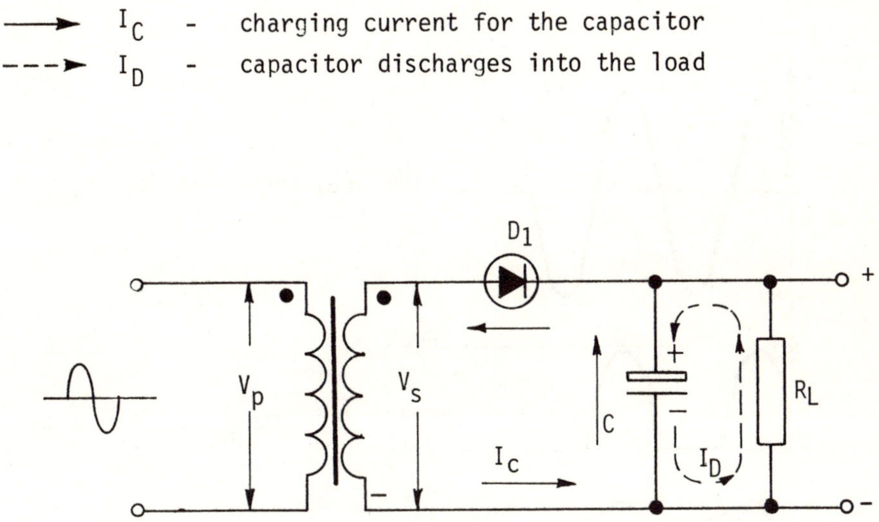

Figure 2.3 Half-wave rectifier with a capacitive filter

to the circuit modifies the circuit operation and output waveform. When the diode conducts, current passes through the load and through C. C charges almost instantly because there is negligible resistance in its charge path. There is only the resistance of the transformer winding. After one-quarter of an input cycle, C is charged to the peak voltage value. During the next quarter cycle, the input is going negative and C tries to discharge. The discharge path is through the load and the time taken for C to discharge is governed by the value of the load. The time constant ($C \times R_L$) is the time taken for the voltage across the capacitor to change by 63 per cent of its former steady value; therefore, the larger the value of R_L, the longer the time constant.

At a first glance it appears that C could discharge through D_1 and the transformer. However, C is charged, and therefore D_1 cathode is at the potential of V_{sm}, and shortly after one quarter cycle of input (90°), D_1 is switched off, i.e. the cathode is more positive than the anode.

The capacitor begins to discharge through the load, and the voltage across it decreases. Before it can discharge completely, the next cycle of input voltage forward biases the diode, and the capacitor recharges to the peak value. Figure 2.4 shows this action graphically.

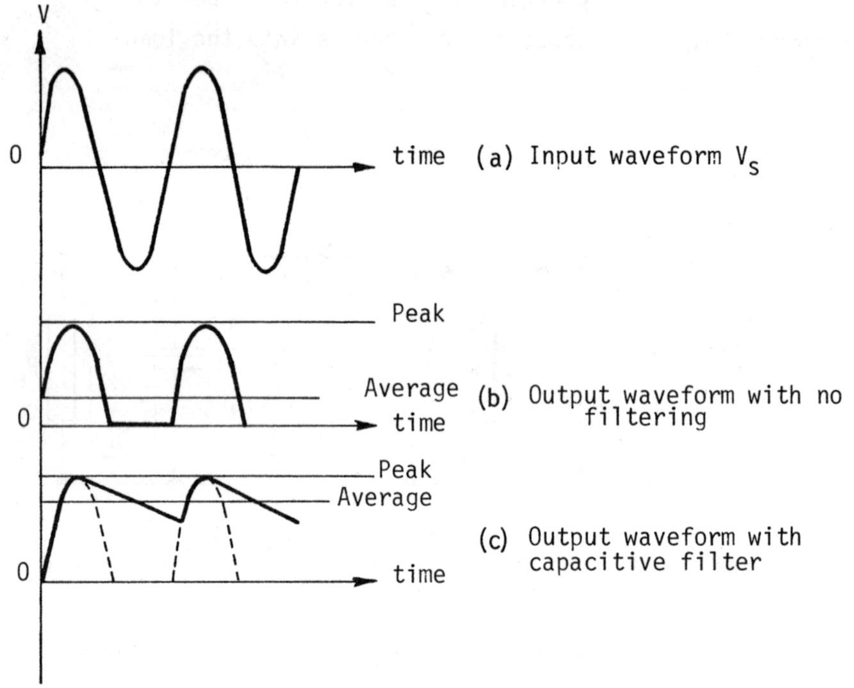

Figure 2.4 Waveforms showing the action of a capacitor as a filter

The heavy line showing the waveform in Figure 2.4(c) represents the voltage across the capacitor, and it shows that not only has the ripple level been reduced, but also that the average level of the waveform has been raised.

If a larger capacitor was used, the time constant is even larger, and C will not discharge as much during the time the diode is off. Therefore, the average level of the output waveform will be even higher. The size of the capacitor also determines the length of time that the diode will conduct. Figure 2.5 shows a comparison of conducting time for a large and a small capacitor. When the diode is not conducting, the capacitor is supplying the current through the load. Therefore, if a high-load current is required, the capacitor needs to be very large. This indicates the limitations of the capacitive filter - it is most useful at low current.

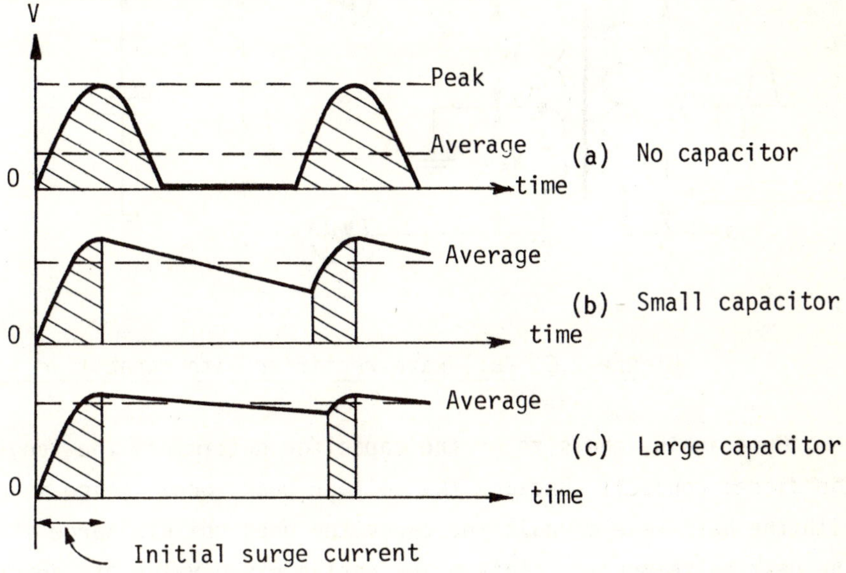

The diode conducts for the shaded portion of the waveforms

Figure 2.5 Comparison of diode conducting times

If the load demand on the circuit is excessive, the capacitor will discharge to a lower voltage level, and the diode will therefore have to conduct longer and harder. A fully discharged capacitor appears as a short circuit to the rectifier, and the initial surge current can be extremely large which could damage the diode. With a large capacitor the diode conducting time is short, but the peak diode current is much higher. It is good practice to have a few ohms in series with the diode to protect it from the initial surge current. The transformer secondary resistance is usually sufficient with modern diodes.

2.4 FULL-WAVE RECTIFIERS WITH CAPACITOR FILTER

A graphical representation of the waveforms obtained from the circuit in

Figure 2.6 is shown in Figure 2.7. It can be seen that the capacitor has the same effect in this circuit as it did in the previous example. It reduces the level of ripple and increases the average level of the output.

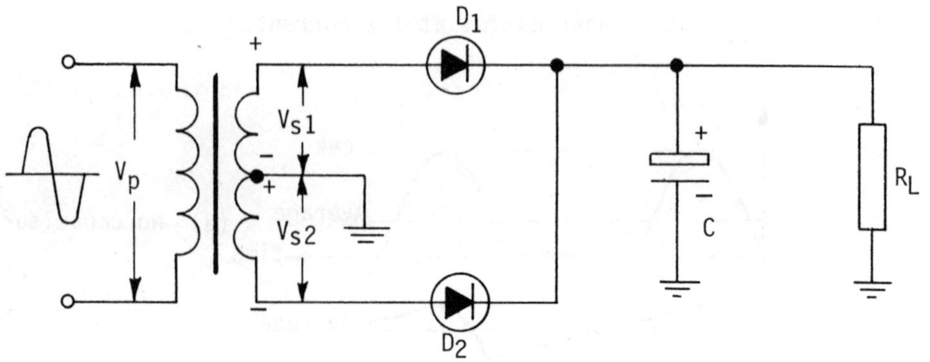

Figure 2.6 Full-wave rectifier with capacitive filter

Once again, the size of the capacitor determines the length of time the diodes conduct. Because the voltage peaks occur twice as often as with the half-wave circuit the capacitor does not discharge very much before the next pulse occurs. This means that a given value of capacitor will filter a full-wave circuit better than it will a half-wave circuit.

Another consideration is the effect of the capacitor on the peak inverse voltage of the diode. From Figure 2.8 it can be seen that with the capacitor charged to the peak of the secondary voltage and with V_S changing polarity (in the next half cycle of input), the diode will have $V_S + V_C$ in series across it in the reverse direction.

This is true for both the full-wave and half-wave rectifier, and therefore, the P.I.V. must be sufficient to handle the sum of V_S and V_C (i.e. twice the peak voltage) when a capacitor filter is used. With a full-wave bridge rectifier circuit, however, this situation does not apply.

As can be seen in Figure 2.9, the diodes are never exposed to more than V_S. D_3 appears to have twice the peak voltage across it, but D_1 is conducting and Point A is effectively at ground potential. D_3, therefore, has only the peak voltage across it. Thus, the P.I.V. rating required for a bridge circuit is less than that for a full-wave circuit.

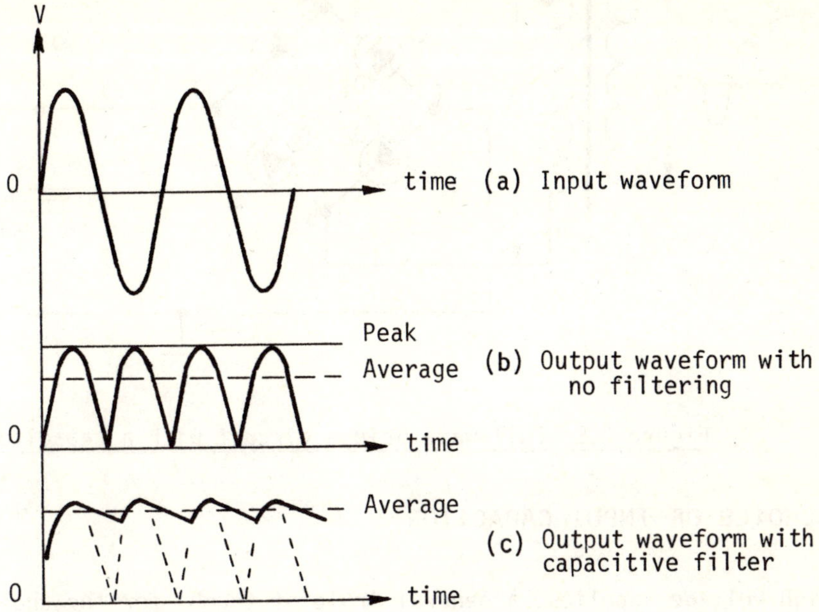

Figure 2.7 Waveforms showing the action of a capacitor as a filter in a full-wave circuit

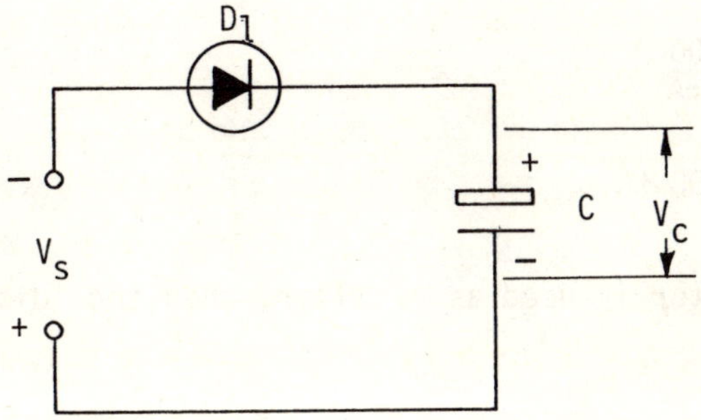

Figure 2.8 D_1 reverse biased with V_s and V_C across it

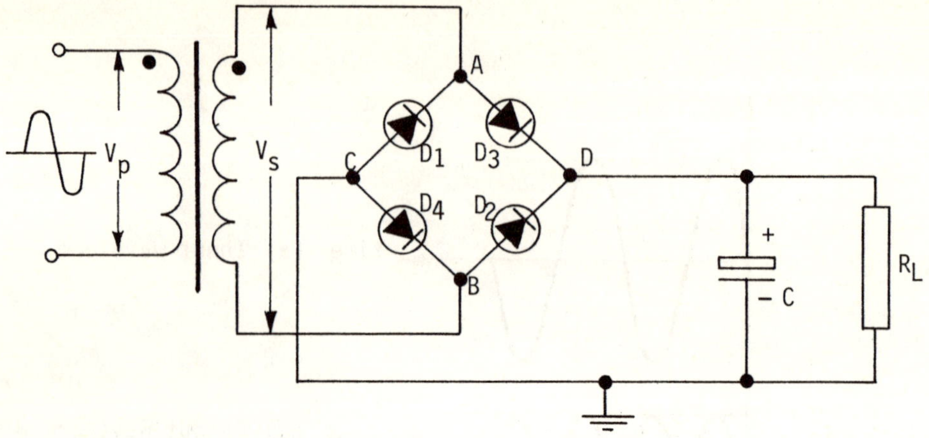

Figure 2.9 Full-wave bridge circuit with a capacitor filter

2.5 CHOICE OF INPUT CAPACITOR

For high voltage supplies, a general 'rule of thumb' for the choice of capacitance value is to use 1.0 μF of capacitance for every milliamp of load current. However, the capacitive reactance at the ripple frequency must be less than the value of R_L.

With an output voltage of 200 V and 1 mA of current through the load, the load resistance will be:

$$R = \frac{V}{I}$$

$$= \frac{200}{1\text{mA}}$$

$$= 200 \text{ k}\Omega \ .$$

If a 1.0 μF capacitor is used as a filter, then the 'discharge' time constant will be:

$$C \times R_L = 1 \times 10^{-6} \times 200 \times 10^3$$

$$= 200 \text{ milliseconds.}$$

To determine the filter input frequency, we must consider the type of rectifier employed. If a full-wave circuit is used on 50 Hz mains voltage, then the frequency is 100 Hz, and the 'period' of input is equal to: 1/frequency, or 10 milliseconds.

The filter capacitor, which delivers energy to the load, has its charge 'topped up' every 10 milliseconds. The time constant of the output is very long compared to the input charge period, and therefore, the capacitor does not discharge very much between input periods due to the high value of load resistance (200 kΩ). However, if the output voltage is reduced to say 5 V and 1 mA of current through the load, then the load resistance equals V/I or 5V/1 mA = 5 kΩ,

With a 1.0 μF capacitor, the 'discharge' time constant will be $C \times R_L$ = 5 milliseconds. The capacitor is still 'topped up' every 10 milliseconds, but the shorter time constant of the output circuit discharges it to a much lower voltage before the next input period. Figure 2.10 shows the current paths of the circuit just described.

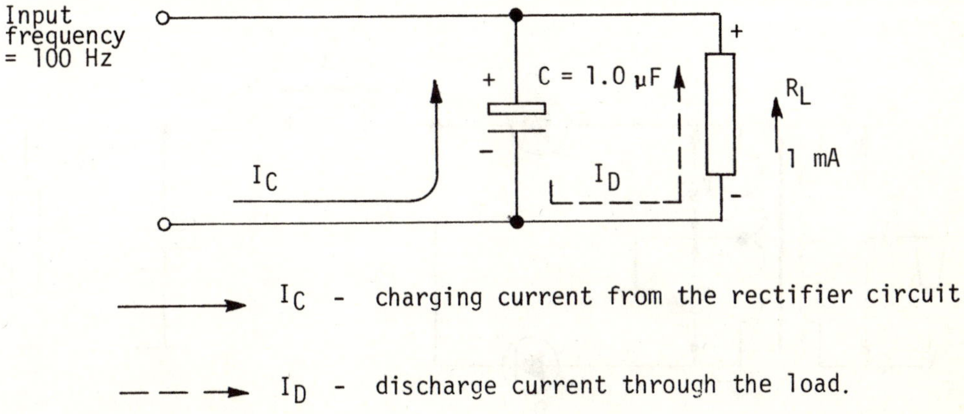

Figure 2.10 Charge and discharge currents through a capacitor filter

From the two examples just discussed, it can be seen that for a low voltage power supply, a larger value of capacitor is required to make the 'discharge' time constant longer. This enables the input circuit to keep the capacitor charged more fully, and hence reduce the amplitude of the ripple voltage.

The discharge time constant must be long compared to the period of the input voltage frequency. The larger the value of the load resistance, the smaller the value of the filter capacitor required for a given ripple, because ripple amplitude is related to the discharge time constant.

A distinct disadvantage of capacitor input filters, including π and CR section (which will be discussed shortly), is that high load currents (low resistance loads) result in short discharge time constants. This means the capacitor discharges to a lower voltage level before the rectifier circuit conducts; the ripple amplitude is therefore increased, and the diode current is greater.

2.6 L - C SECTION FILTERS

A further improvement to a filtering circuit can be made by adding an inductor between the diode output and the capacitor. This type of circuit is called an 'L section' because it resembles an inverted 'L'. A schematic diagram of this type of filter is shown in Figure 2.11.

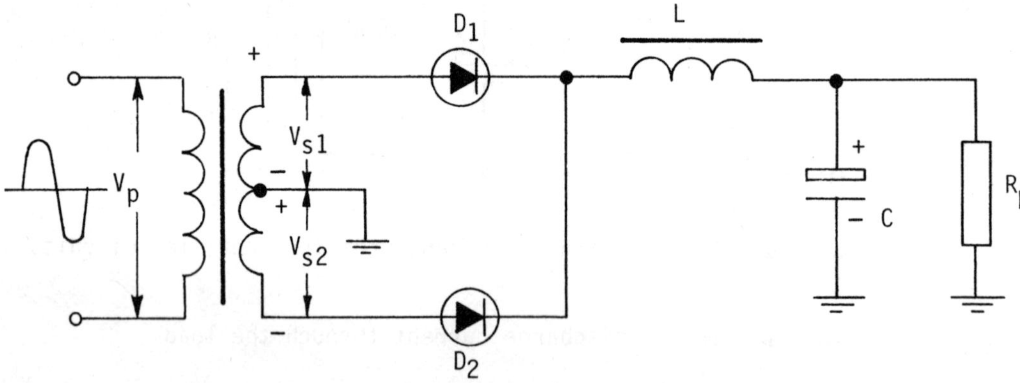

Figure 2.11 L - C Section Filter

The inductor, or choke, opposes changes in current, and energy is stored in L in the form of a magnetic field. During the interval between pulses, energy comes out of L as the field collapses. If a half-wave circuit was used with a L-C section filter, energy would be released from L during the second half cycle of input. The counter e.m.f. set up by the changing current will tend to oppose the change and keep the current constant. The result is a smoother D.C. voltage, which is then used to charge the filter capacitor. An inductor offers a high impedence to A.C. and low impedance to D.C. The action to a L-C filter is shown in Figure 2.12.

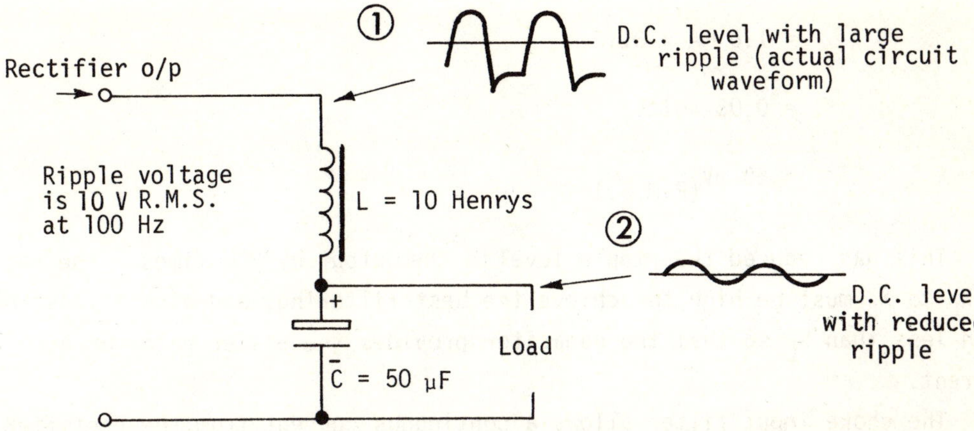

1. Waveform at the input to the filter

2. Waveform expected across the capacitor

Figure 2.12 L-C Section Filter operation

With 10 V R.M.S. of ripple at 100 Hz the L-C section behaves as a voltage divider.

$$X_L = 2\pi fL$$

$$= 6280\ \Omega$$

$$X_C = \frac{1}{2\pi fC}$$

$$= 31.8\ \Omega$$

The impedance of the circuit (Z) equals 6280 - 31.8 Ω = 6248.2 Ω, and therefore the current is:

$$I = \frac{V}{Z} = \frac{10}{6248.2} = 0.0016 \text{ amperes}$$

$$= 1.6 \text{ mA}$$

With 1.6 mA of current in the circuit, we end up with a ripple voltage across C approximately equal to I × X_C.

$$V_C = IX_C$$

$$= 1.6 \text{ mA} \times 31.8 \ \Omega$$

$$= 0.05 \text{ volts}$$

$$= 50 \text{ mV}_{(R.M.S.)}$$

This has reduced the ripple level in the output by 200 times. The ratio of X_L to X_C must be high to achieve the best filtering, and also X_C must be much less than R_L so that the capacitor provides the easier path for A.C. current.

The choke input filter allows a continuous current from the rectifier diodes, rather than the pulsating current demanded by the capacitor input filter, and because of this, it has better voltage regulation. Maximum current is available to the load with this type of operation, so this filter is more suited to higher load-current demand.

On the debit side, the output voltage of the filter is equal to the average of the A.C. voltage delivered to the rectifier, rather than the peak value. It is also very costly, because a physically large, low resistance inductance is required for good regulation.

If a high degree of filtering is required with a high current ability, then two L-C section filters may be used in series. The level of the ripple on the D.C. output can be further reduced by including a capacitor to ground prior to the inductor. This type of filter circuit is shown in Figure 2.13 and is called a Pi-filter (π).

Capacitor C_1 performs the same function as the C input filter, and is selected to have a low reactance X_C at the ripple frequency. The charge stored in C_1 is then released into a L-C section filter L and C_2. The output voltage of this type of filter is very nearly pure D.C.

Although a π-filter is a more efficient filter, C_1 is still connected across the rectifier output, and high pulses of current are required to keep it charged. If the load demand is too great, the high-current pulses may damage the diodes. The π-filter is used on medium-current equipment;

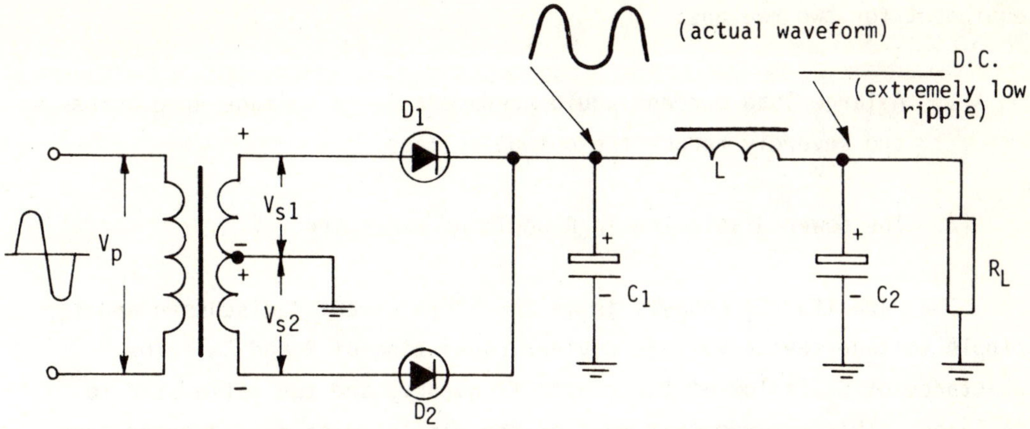

Figure 2.13 Pi-filter

it does produce a higher output voltage than the L-C filter, but the regulation is not as good.

2.7 CR-SECTION FILTER

The inductor in either of the previous filters may be replaced by a resistor to form a CR-section filter. This type of filter is less expensive. A full-wave bridge with a CR filter is shown in Figure 2.14. The filtering action

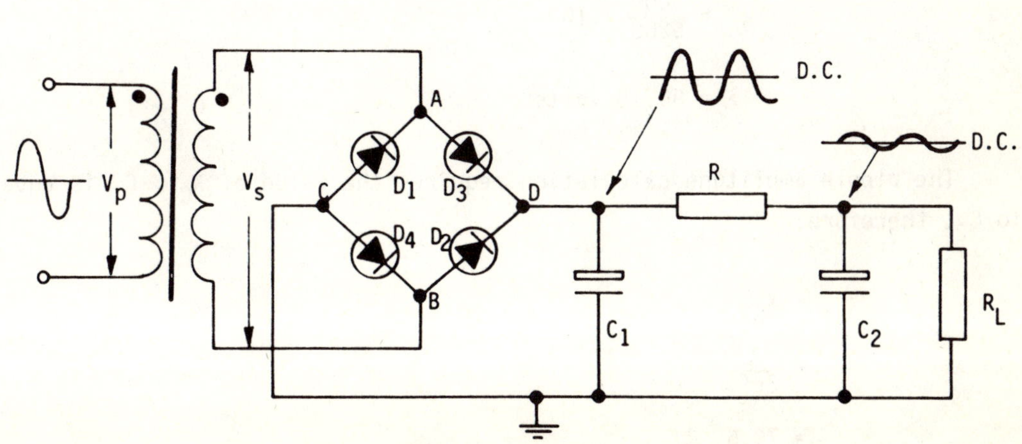

Figure 2.14 CR-Section Filter

is not as good, and there is some loss of D.C. voltage at the output because of the voltage drop across the resistor. It is limited to low-current equipment for two reasons:

1. A large load current would produce a large voltage drop across R and severely reduce the output voltage:

2. The power dissipated in R would be excessive.

The capacitor C_1 behaves as in the other circuits discussed and the ripple voltage sees a voltage divider consisting of R and C_2. The reactance of C_2 is low at the ripple frequency, and the ratio of R to X_C is large. This ensures that most of the ripple voltage is dropped across R and very little is present across C_2 and the load.

If the value of the components in Figure 2.14 are known, the expected D.C. output voltage and ripple amplitude can be calculated. Assume $C_1 = C_2 = 20\ \mu F$, $R = 200\ \Omega$, and $R_L = 5000\ \Omega$. The input voltage is 100 volts D.C. with 10 volts (R.M.S.) of A.C. ripple. The D.C. output voltage is determined by R and R_L acting as a voltage divider. Therefore:

$$\text{D.C. } V_{out} = \frac{R_L}{R + R_L} \times V_{in(D.C.)}$$

$$= \frac{5000}{200 + 5000} \times 100$$

$$= \frac{5000}{5200} \times 100$$

$$= 96.15 \text{ volts}$$

The ripple amplitude calculation requires the value of X_C. C_1 is equal to C_2, therefore:

$$X_{C1} = X_{C2}$$

$$= \frac{1}{2\pi fC}$$

$$= 79.6\ \Omega.$$

The value of X_C is about one-third the value of R and therefore must be taken into account.

The A.C. ripple voltage equals:

$$\frac{X_C}{\sqrt{R^2 + X_C^2}} \times V_{in(R.M.S.)}$$

$$= \frac{79.6}{\sqrt{200^2 + 79.6^2}} \times 10$$

$$= \frac{79.6}{215.25} \times 10$$

$$= 3.69 \text{ volts.}$$

If X_C were much smaller than R, i.e. a larger value capacitor, the amplitude of the ripple at the output would be considerably reduced. If $C_2 = 50\ \mu F$, then:

$$X_{C2} = \frac{1}{2\pi fC}$$

$$= 31.8\ \Omega.$$

Therefore the A.C. ripple is equal to:

$$\frac{31.8}{\sqrt{200^2 + 31.8^2}} \times 10$$

$$= \frac{31.8}{202.5} \times 10$$

$$= 1.57 \text{ volts.}$$

Generally, R must be much less than R_L (to maintain the level of D.C. output voltage), X_{C2} must be much less than R (to reduce the level of A.C. ripple in the output), and X_{C2} must be one-fifth the value of R_L, so that the capacitor provides the low-resistance path for the A.C. current.

2.8 VOLTAGE REGULATION

Voltage regulation is a measure of a circuit's ability to maintain a constant output voltage under varying load conditions.

If a power supply delivers 100 volts with no load connected, and this output drops to 80 volts when the load is connected, then the regulation of the power supply can be expressed as a percentage:

$$\text{Percentage regulation} = \frac{V_{NL} - V_{FL}}{V_{FL}} \times 100$$

where V_{NL} = no-load voltage

V_{FL} = full-load output voltage

In the case given above:

$$\text{Percentage regulation} = \frac{100 - 80}{80} \times 100$$

$$= \frac{20}{80} \times 100$$

$$= 25 \text{ per cent}$$

The lower the percentage, the better the regulation.

If the regulation of a simple rectifier and filter circuit is poor, it may be improved by:

1. Using full-wave rectification.

2. Ensuring minimum resistance in both transformer and filter choke.

3. Using L.C. filter if the current is high.

4. Ensuring solid-state diodes are used; they have a low voltage drop across them.

2.9 FAULTS IN FILTER CIRCUITS

Two main faults can be experienced with the inductor in a filter circuit, and they are:

1. Open circuit coil

2. Shorted turn in the coil

The most common one is an open circuit, and the filter choke is tested in the same manner as a transformer primary or secondary.

The capacitor on the other hand can present three unserviceable conditions:

1. Open circuit

2. Short circuit

3. Leaky

If it is a case of short circuit or open circuit, the capacitor is unserviceable because it cannot store charge. A leaky capacitor has a weakened dielectric, and it therefore has a lower resistance than normal.

A further point to bear in mind is a 'change' in capacitance value over time. This is especially noticeable with electrolytic capacitors. As they age, they dry out and their capacitance tends to decrease. This leads to an increase in capacitive reactance. They then offer more resistance to A.C., and the circuit operation is affected as a result.

If an electrolytic capacitor changes value in a filtering circuit, the reactance ratio is changed (see Figure 2.11) and the ripple amplitude increases. This can cause such faults as hum in audio equipment, and general instability in operation. Electrolytic capacitors can be tested with an ohmmeter set to the R × 1000 range, as shown in Figure 2.15.

With a good capacitor the needle moves quickly towards the low resistance side of the scale as the capacitor charges (Figure 2.15a) and then it slowly drops back towards infinity. The final resting point of the needle indicates the insulation resistance of the capacitor (Figure 2.15b). The insulation resistance of an electrolytic capacitor varies with the size,

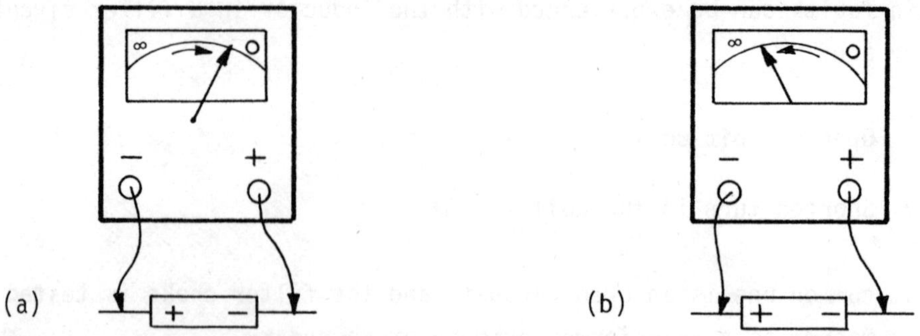

Figure 2.15 Testing a capacitor with an ohmmeter

but approximately 1 megohm and upwards is common.

If the capacitor has a short-circuit condition, the ohmmeter needle will go to zero and stay there. If the capacitor is open circuit, then there will not be any charging action and the ohmmeter needle will read a very high resistance. However, this fault should be double-checked by reversing the meter leads and checking in the opposite direction.

2.10 SELF-EVALUATION QUESTIONS (ANSWERS IN APPENDIX A)

1. What is the purpose of filtering?

2. A 240 V D.C. voltage has a 2 volt peak-to-peak ripple voltage present on it. What is the percentage ripple?

3. The regular capacitor in a C-input filter is replaced with a capacitor of a larger value. What effect does this have on:

   (a) The average output level?

   (b) The diode current?

4. Which diode specification should be considered when using a C input filter arrangement?

5. A high-voltage power supply has an unloaded output of 350 V. If the loaded output is 330 V, what is the percentage regulation?

6. If high current was the prime consideration of a power supply rather than voltage, what type of filter would you use?

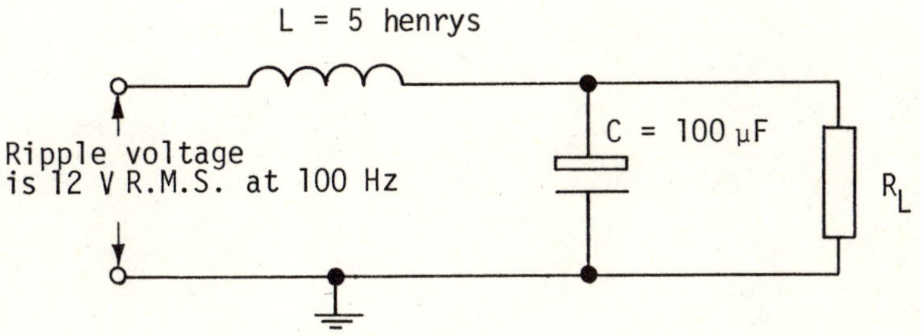

Figure 2.16

7. From the circuit shown in Figure 2.16, calculate:

   (a) X_L

   (b) Filter current

   (c) Peak output ripple voltage

8. List three practical ways to improve the regulation of a basic rectifier/filter circuit.

9. What is the ripple frequency of:

   (a) A full-wave circuit operating at 100 Hz?

   (b) A half-wave circuit operating at 400 Hz?

3 Voltage Regulators

3.1 INTRODUCTION

The need for some sort of voltage regulation can be seen by examining the simulated regulator in Figure 3.1. With 20 V from the rectifier and filter

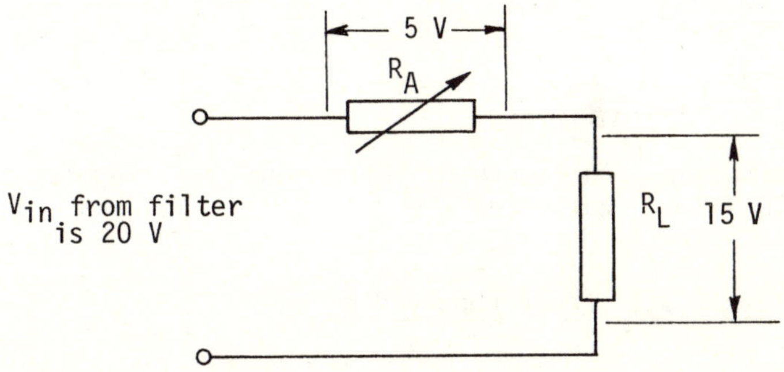

Figure 3.1 Simulated voltage regulator

circuit, 15 V is required across the load and 5 V is dropped across the variable resistance R_A. If the input voltage suddenly increases, the voltage drop across R_A would have to increase to maintain the voltage across R_L constant. That is, the resistance of R_A would have to be increased. On the other hand, a decrease in the value of R_L (more current required by the load) would increase the load current and therefore increase the voltage across R_A. To maintain the output voltage constant, R_A would have to decrease in value. An increase in the value of R_L would cause less current through the load. The voltage drop across R_A would decrease, and the value of R_A would have to be increased to hold the voltage across R_L at 15 volts.

It can be seen from these examples that the output voltage can be held constant by varying R_A to suit the circuit conditions. To keep adjusting R_A manually would be impossible, so a device called a voltage regulator is used to perform the task electronically. Figure 3.2 shows the format of the final power-supply circuit.

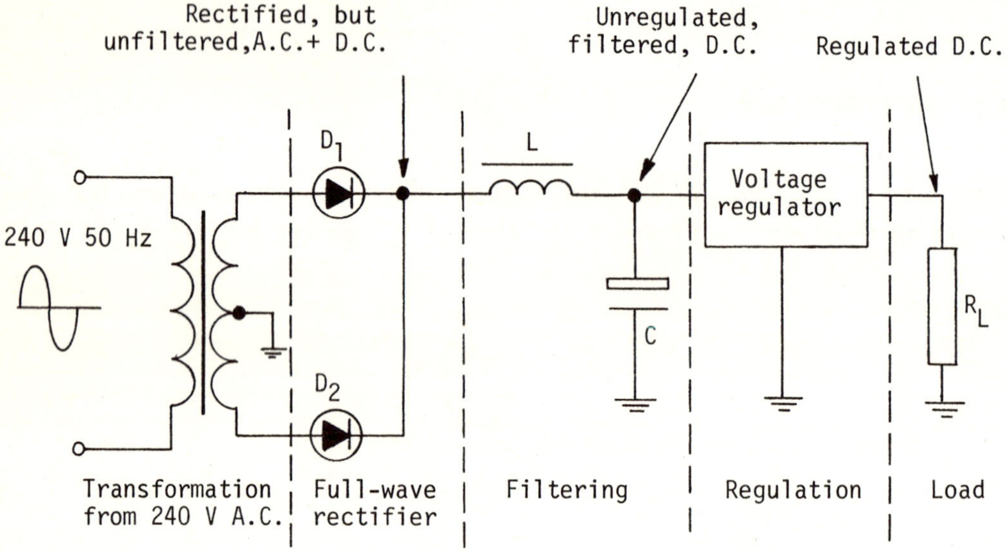

Figure 3.2 Power supply diagram

3.2 ZENER DIODE REGULATOR

The simplest form of voltage regulator circuit is the zener diode regulator (see Figure 3.3). A zener diode works in a 'reverse breakdown mode' with the cathode connected to the positive terminal of the supply. When the diode is in breakdown, its voltage remains fairly close to the breakdown value, even if the current varies. The reverse characteristic curve for a zener diode is shown in Figure 3.4.

If the unregulated D.C. voltage input increases, the current through R_S also increases. This extra current is diverted through the diode instead

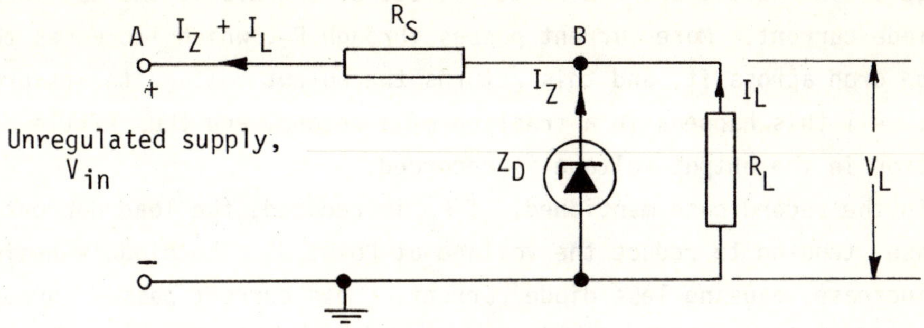

Figure 3.3 Zener diode shunt regulator

of passing through the load. The diode voltage is virtually unaffected by the increase in current, and the load voltage therefore remains fairly constant. If more current is required by the load (i.e. R_L is reduced), then the zener diode circuit allows the extra current to be supplied without affecting the load voltage.

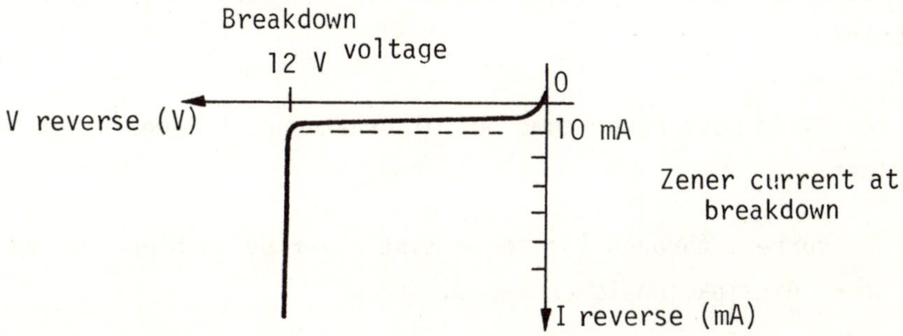

Figure 3.4 Reverse characteristic for a zener diode

Looking more closely at the operation of this circuit, if the input voltage at Point A (in Figure 3.3) increases, the voltage at Point B will tend to increase also. This increases the diode reverse voltage. The zener diode's dynamic resistance, which is inversely proportional to the reverse voltage across it, therefore decreases. This increases the zener current through R_S, hence increasing the voltage drop across R_S and maintaining the output voltage level constant.

In summary, if the input voltage increases, then the reverse zener voltage is increased, and the diode resistance decreases, thereby increasing the diode current. More current passes through R_S, which increases the voltage drop across it, and this returns the output voltage to its previous value. All this happens in a fraction of a second, and thus little variation in the output voltage is recorded.

In the second case mentioned, if R_L is reduced, the load current will increase, tending to reduce the voltage at Point B. The diode's resistance will increase, causing less diode current. Less current passes through R_S, and therefore less voltage is dropped across it. This restores the voltage at Point B, i.e. the output voltage is back to normal. In summary, when $R_L \downarrow$ then $I_L \uparrow$, $V_{RL} \downarrow$ and $R_Z \uparrow$, therefore $I_Z \downarrow$, $V_{RS} \downarrow$, hence V_{out} remains constant. Since the regulator and the load are in parallel, the circuit is called a shunt regulator.

It is important to realize that the description of the operation of voltage regulators has been time expanded to show exactly what happens. In reality, a change in the output condition is barely noticeable, and in complex regulators it is not noticeable at all.

When considering zener diode regulators, two important factors must not be overlooked:

1. The diode must never come out of breakdown, or regulation will be lost.

2. The current through the zener must never be so high that maximum power dissipation is exceeded.

These factors are taken into consideration when calculating the value of the series resistor R_S.

3.2.1 CALCULATION OF R_S

Assume the following values for the circuit shown in Figure 3.3:

V_{in}	= 20 volts
Zener voltage	= 12 volts
Load current (I_L)	= 55 mA
Load current variation	= ± 5 mA (I_L varies between 50 and 60 mA)
Zener current	= 10 mA (Zener current will increase as the load current decreases)

The maximum current (I_{max}) through R_S is $I_Z + I_L$ = 60 + 10 mA = 70 mA. Then R_S will drop 70 mA × R_S volts, and this must be equal to $V_{in} - V_{out}$ (V_{out} = zener voltage). Therefore:

$$R_S = \frac{V_{in} - V_{out}}{I\ max}.$$

For our particular zener:

$$R_S = \frac{20 - 12}{70 \times 10^{-3}}$$

$$= \frac{8}{0.07}$$

$$= 114\ \Omega$$

The power dissipated by R_S is equal to $(I_{max})^2 \times R_S$.

$$\text{Power} = (I_{max})^2 \times R_S$$

$$= 0.07^2 \times 114$$

$$= 559 \text{ milliwatts}$$

The power dissipated by the zener diode when no current is being drawn by the load, can be calculated. With no load current being drawn, $I_Z = I_{RS}$.

Therefore:

$$I_{RS} = \frac{V_{in} - V_L}{R_S}$$

$$= \frac{20 - 12}{114}$$

$$= 70 \text{ mA}$$

$$\text{Power dissipated} = 12 \times 70 \text{ mA}$$

$$= 840 \text{ milliwatts.}$$

If the zener diode characteristics are not known, a general rule of thumb for calculating the value of R_S is to regard I_{max} as 1.5 times the average load current. Using this rule for the previous example, we have I = 50 mA, V_{in} = 20 V and V_Z = 12 V. Then:

$$R_S = \frac{20 - 12}{1.5 \times 50 \text{ mA}}$$

$$= \frac{8}{75 \times 10^{-3}}$$

$$= 106\ \Omega$$

This is very close to the value calculated with all the circuit information known, and it gives a slight increase in the maximum power to be dissipated by the zener.

$$\text{Power dissipated} = 12 \times 75 \text{ mA}$$

$$= 900 \text{ milliwatts}$$

3.2.2 DISADVANTAGES

A disadvantage of a zener diode regulator circuit is that the output voltage varies if the load currentis subject to *considerable* variations. Zener diodes are also very sensitive to temperature changes, and in this case also, the output will vary. The power losses in R_S and in the zener diode itself are also quite high when high current is supplied, and this reduces the circuit's efficiency.

3.3 SIMPLE TRANSISTOR SHUNT REGULATOR

A simple shunt voltage regulator with a transistor in parallel with the load is shown in Figure 3.5. If the load resistance increases, the output voltage would tend to rise, i.e. become more positive, raising the base voltage and hence the forward bias on Q_1. With more forward bias (V_{BE}) the collector current increases, increasing the voltage drop across R_1, and the output voltage is reduced to its former value.

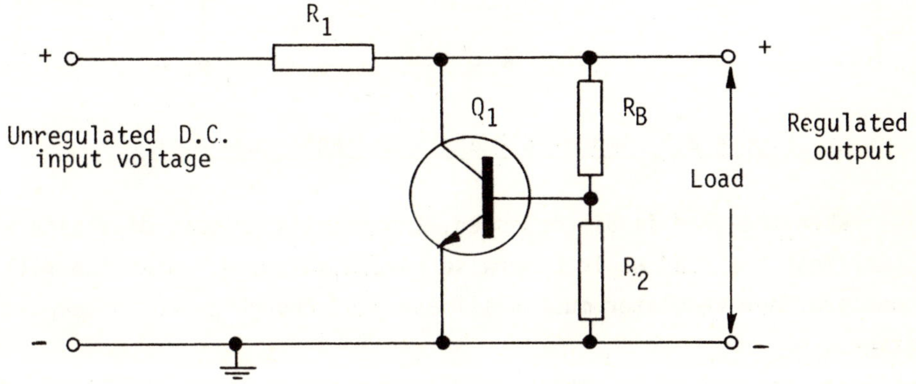

Figure 3.5 Transistor shunt regulator

Figure 3.6 shows an improvement on the simple circuit. The zener diode maintains a constant voltage between base and collector of Q_1 which is equal to its breakdown voltage.

If the load resistance increases, the output voltage tends to become more positive. There will be more voltage across the Z_D/R_S leg, and the zener current will increase, increasing the voltage drop across R_S. This will increase the forward bias on Q_1, and hence Q_2. I_C of Q_2 will increase, and therefore more voltage will be dropped across R_1, returning the output voltage back to its former value.

When the load resistance decreases, the circuit operates in reverse, but V_{CB} of Q_1 always equals the zener diode voltage. The output voltage falls, I_Z will be less, this will make V_{R_S} less, and therefore, Q_1 forward bias will be reduced. Less I_C flows through R_1 reducing the voltage drop across it and restoring the output voltage.

The problem with shunt regulators is the waste of power in the regulator

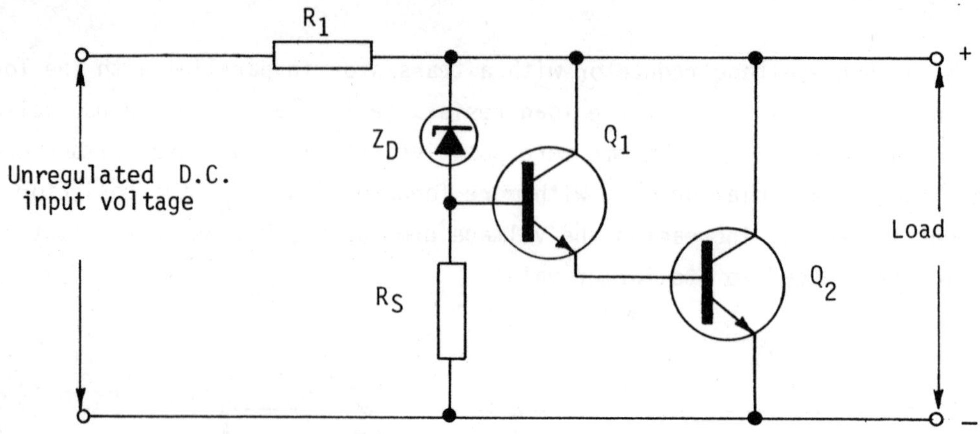

Figure 3.6 Improved transistor shunt regulator

itself. When the load is disconnected, the transistor must dissipate all the power that the load would otherwise have dissipated. When the full load is connected, the regulator must dissipate just enough power to keep it operating.

All the heating of R_1, which is present regardless of the load condition, is also a waste of power. Shunt regulators are usually only used when a small current is to be controlled. One advantage of the shunt regulator is its ability to withstand a short circuit without being damaged; the only consequence is that the series resistor gets hot.

3.4 SERIES TRANSISTOR REGULATORS

As shown in Figure 3.1, an automatically varying impedance in series with the power supply can handle all of the circuit demands for current. A series transistor regulator does just that, and it can vary from a simple to a highly complex circuit. Figure 3.7 shows the circuit diagram of a simple series regulator.

The resistor R_S is chosen to allow the necessary reverse current for the zener diode to operate just in the breakdown region. This ensures that the base Q_1 is *fixed* at the zener diode voltage (in this case 12 V). The regulated output voltage will be slightly less than the zener voltage: i.e. $V_{ZD} - V_{BE}$. The average V_{BE} = 600 mV, and therefore the output voltage will be 11.4 volts. The difference between the input voltage and

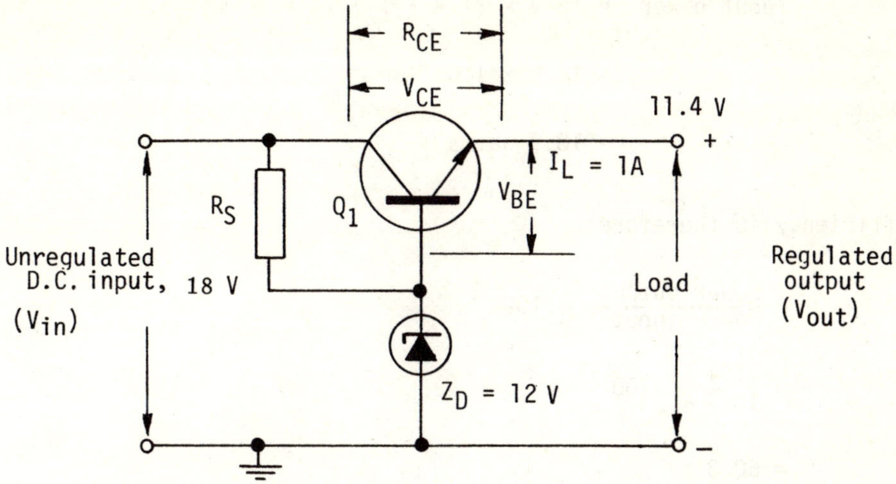

Figure 3.7 Simple series transistor regulator

the output voltage is dropped across the transistor, V_{CE}. Therefore:

$$V_{CE} = V_{in} - V_{out}$$

$$= 18 - 11.4$$

$$= 6.6 \text{ volts}$$

For a 1 A load current, transistor Q_1 must dissipate power, the value of which is found in the following way:

V_{CE} = 6.6 volts

I_C = 1 ampere

Power = 6.6 × 1

= 6.6 watts.

This is wasted power which reduces the efficiency of the power supply.

Output power = 11.4 × 1 A = 11.4 watts

Input power = 18 V × (1 A + I_Z) I_Z = 50 mA

= 18 V × 1.05 A

= 18.9 watts

The efficiency is therefore:

$$= \frac{\text{Power output}}{\text{Power input}} \times 100$$

$$= \frac{11.4}{18.9} \times 100$$

$$= 60.3\ \%$$

This low level of efficiency is one of the drawbacks with this type of circuit. Another problem is the heat dissipation in the series pass transistor Q_1. A high operating temperature can severely degrade the transistor's reliability.

Operation. If the load resistance in the circuit shown in Figure 3.7 decreases the output voltage will tend to fall, i.e. become less positive. This means that V_E of Q_1 is reduced, and the emitter potential moves away from the fixed base potential. Therefore, V_{BE} increases.

When the forward bias of a transistor is increased the junction resistance (R_{CE}) decreases. The load current is through R_{CE}; therefore less voltage is dropped across the transistor, and the output voltage returns to its former value.

When the load current decreases, the output voltage will tend to increase towards the base voltage. This reduces V_{BE}, which causes R_{CE} to increase. With more resistance, V_{CE} increases, and the output voltage returns to normal.

Similarly, if there is a sudden change in the unregulated input voltage, the regulator will respond in a way that will maintain the output voltage at its predetermined level. For example, if the input voltage increased, the output voltage would tend to increase also. This would bring the emitter voltage of Q_1 closer to the base voltage, thereby reducing V_{BE}. The transistor's resistance would increase, dropping more voltage across Q_1, and the output would return to normal.

3.4.1 FEEDBACK REGULATORS

The circuit in Figure 3.7 is called an 'open-loop' type, because there is no way to ensure that the output voltage will stay *absolutely constant* under varying operational conditions, such as temperature variation and component tolerances. Also, it always takes some error signal to initiate the action of the regulator. If the error signal is amplified, the circuit becomes more sensitive to voltage variations, and this results in better voltage regulation. It also makes the output voltage adjustable. A block diagram of such a system is shown in Figure 3.8.

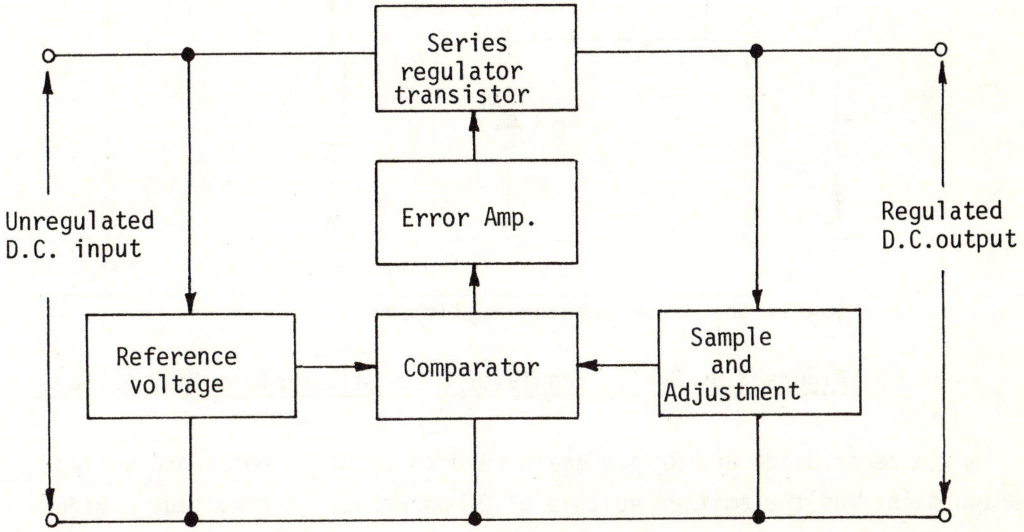

Figure 3.8 Block diagram of a series regulator with feedback

The regulator in Figure 3.8 is a 'closed loop' type, in which the base current of a series pass transistor is altered to compensate for any changes in the output voltage. A portion of the regulator output voltage is fed to a comparator along with a fixed D.C. reference. These two voltages are compared to each other. Any difference between the feedback and the reference voltages (the error signal) is used to bias an amplifier (known as an error amplifier) which increases the signal's amplitude. The amplified error signal is then fed to the base of the series transistor to adjust its forward bias and hence its conductivity. A practical circuit which applies this principle is shown in Figure 3.9.

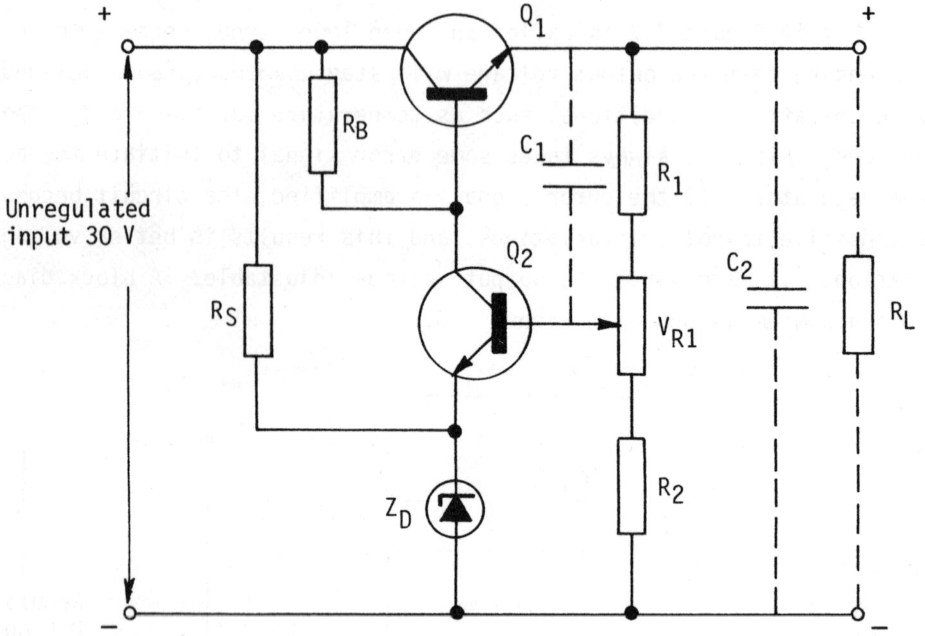

(Q_2 is the error sensing amplifier)

Figure 3.9 Series regulator with adjustable output voltage

The zener diode and R_S are again used to supply a reference voltage which maintains the emitter voltage of Q_2 constant. The voltage divider R_1, V_{R1}, and R_2 across the output senses the value of the regulated voltage and feeds a suitable proportion of it to the base of Q_2. If the zener voltage is 12 volts, a suitable potential for Q_2 base to initiate conduction would be approximately 12.6 V. Resistor R_B puts transistor Q_1 into forward bias, and the second resistor (for the voltage divider bias) is made up of Q_2 and the zener diode. The equivalent circuit is shown in Figure 3.10. V_R is an electronically controlled variable resistance.

Operation. Any change in the resistance of Q_2 will change the bias on Q_1. In Figure 3.9, if V_{R1} is set to give a small value of collector current in Q_2, then the forward bias of Q_1 will be reduced to a state of conduction that may produce perhaps 15 V across the load.

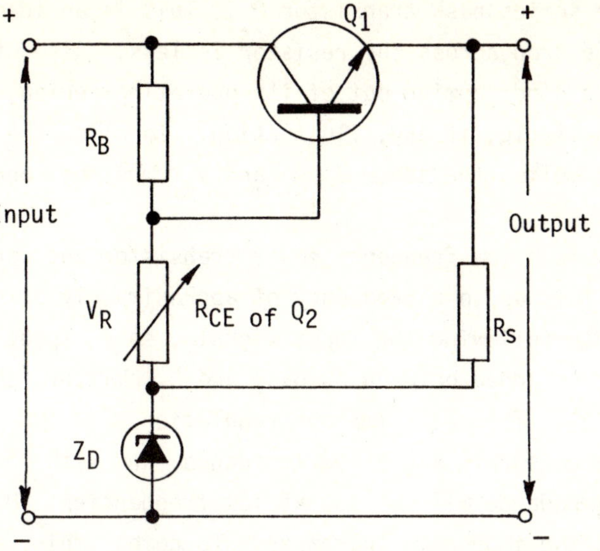

V_R represents the C-E resistance of transistor Q_2,
and R_S is taken to the output

Figure 3.10 Equivalent circuit of a variable output regulator

If the load resistance decreases, the emitter voltage of Q_1 tends to decrease. The reduction in load voltage also reduces the forward bias on Q_2, increasing its resistance. The collector voltage of Q_2 will rise, increasing the forward bias on Q_1, and therefore decreasing R_{CE} of Q_1. With the resistance decreased, less voltage is dropped across Q_1, and the output voltage is restored to its previous level.

On the other hand, if the load resistance is increased, and there is less requirement for current, the emitter voltage of Q_1 will tend to increase. This will increase the forward bias on Q_2, reduce its resistance, and the collector voltage will decrease. The base voltage on Q_1 is also reduced, and hence its forward bias is reduced. With less forward bias on Q_1, R_{CE} is increased, more voltage is dropped across the transistor, and the output voltage is returned to its pre-determined value.

The output voltage is adjustable over a range of voltages from slightly above the zener diode value to about three-fourths of the unregulated input voltage. This is done by adjusting V_{RI}. In Figure 3.10, the bias resistor for the zener diode (R_S) is taken from the lower voltage on the output side of the series pass transistor Q_1. This is an ideal situation, because the voltage drop across the resistor is less. Also, there is less chance of the zener diode coming out of its operating region, because the output voltage has little, if any, fluctuation. To allow the variation to start from zero volts, the zener diode and V_{RI} must be connected to a negative supply.

Transistor Q_1 is a low frequency power transistor and can follow variations in the load up to a frequency of approximately 8 kHz. Above this, it has trouble following the input signals, so a 'speed-up' capacitor (C_1) can be added between Q_2 base and Q_1 emitter (shown in dashed lines in Figure 3.9). This will improve regulation.

A well regulated supply has a low impedance, and with low frequency transistors, the impedance will be low at low frequencies. However, at higher frequencies the impedance increases. To combat this, a capacitor C_2, whose reactance decreases as frequency increases, is connected across the output. The result is a reasonably well regulated power supply from D.C. to high frequencies, with a low output impedance and a ripple of perhaps 5 mV.

The ability of the regulator to oppose any changes in its output voltage can be further utilized to perform electronic smoothing. Any residual mains ripple on the output of the regulator is fed back to Q_2 base via R_1 and C_1, shown in heavy lines in Figure 3.11. Transistor Q_2 sees the ripple as a fluctuation in the output, and the circuit takes corrective action resulting in a much smoother output voltage.

3.5 OVERLOAD PROTECTION

Several methods are available to protect a power supply from burn-out due to excessively heavy current. A fuse in series with the transformer primary, secondary, or in the D.C. output circuit can be used, but these require replacement when blown. Overload relays and thermal relays that trip when there is excessive current can be inserted in the D.C. line, but these require manual resetting. An automatic electronic overload protection circuit is shown in Figure 3.12.

In this circuit, Q_1 is a silicon transistor, $V_{BE} = 0.6$ V. Two silicon diodes and a 1 ohm resistor are connected between the base and emitter of

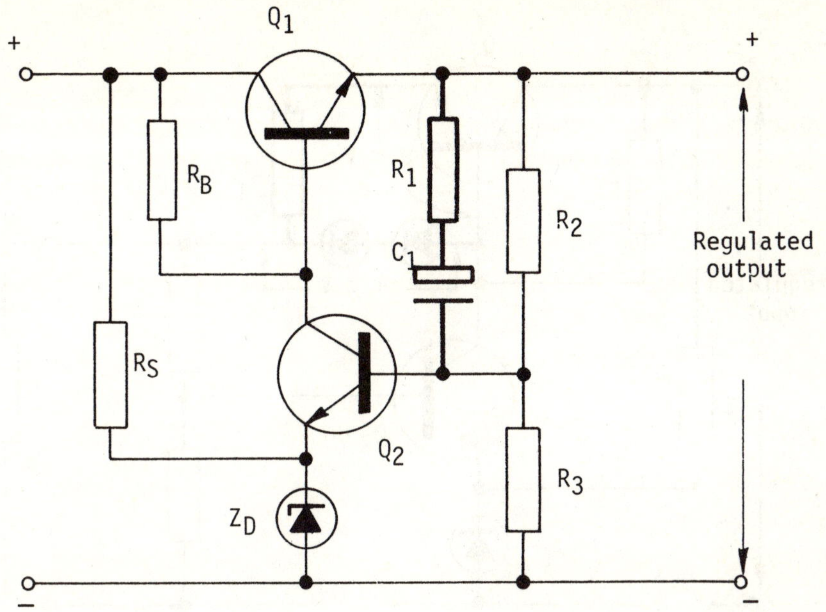

Figure 3.11 R_1 and C_1 provide electronic smoothing

the current passing transistor. The load is connected to the junction of the diodes and the resistor, ensuring that the load current passes through the 1 ohm resistor.

The diodes will not conduct until the voltage drop across them exceeds 1.2 V. With V_{BE} of Q_1 equal to 0.6 V, the load current must produce a 0.6 V drop across R_1.

If the load is drawing 500 mA, the voltage drop across R_1 is $1 \times 500 \times 10^{-3} = 0.5$ V, the diodes will not conduct. As the load increases the current also increases, and the voltage drop across R_1 exceeds 0.6 V causing the diodes to conduct. As the diode current increases, the base current of Q_1 decreases. This in turn reduces the collector current of Q_1. Transistor Q_1 will now only pass a small value of current, and even if the output is short circuited, there is insufficient current through Q_1 to damage it.

A similar method of limiting current to prevent damage to the current passing transistor involves replacing the diodes with a transistor. This is shown in Figure 3.13. When an excessive current passes through the current sensing resistor R_1, the transistor Q_3 turns on. This decreases the forward

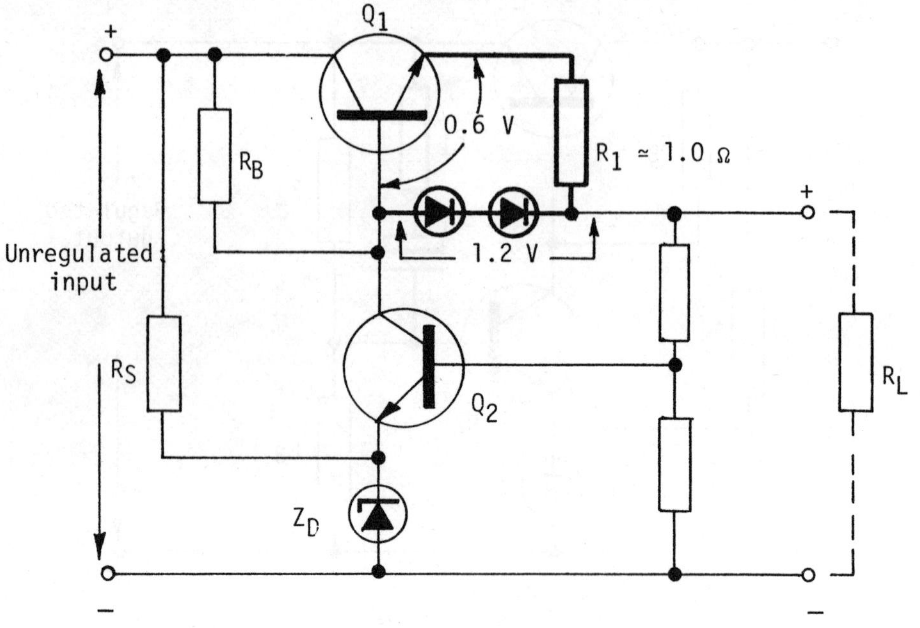

Figure 3.12 Overload protection circuit shown in heavy lines

bias voltage on Q_1 and limits the current flow through it.

3.6 OPERATIONAL AMPLIFIERS IN REGULATORS

Monolithic I.C. op-amps, which are specifically designed for voltage regulation, can be found in better quality voltage regulator circuits. They retain the series current passing transistor, but the error amplifier takes the form of an operational amplifier. Such a circuit configuration is shown in Figure 3.14.

The sample of the output voltage is obtained from voltage divider R_2/R_3. The ratio can be adjusted to provide the required output voltage level; or, alternatively, an adjustable leg can be used as in Figure 3.9. R_S and the zener diode provide the reference voltage, and R_1 limits the input current to the op-amp.

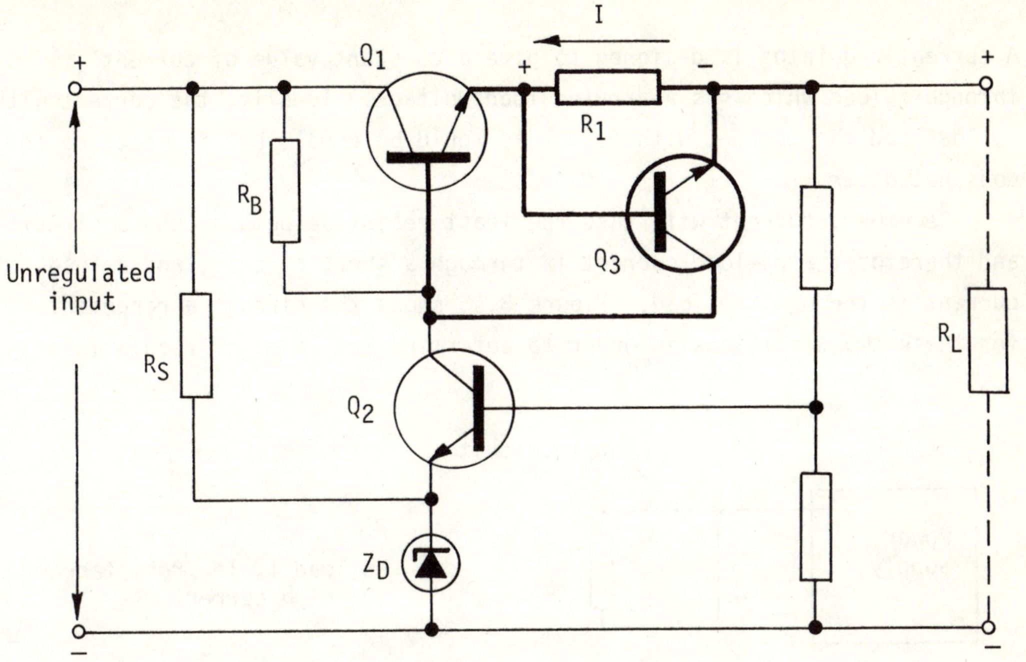

Figure 3.13 Transistor regulator with controlled current-limiting

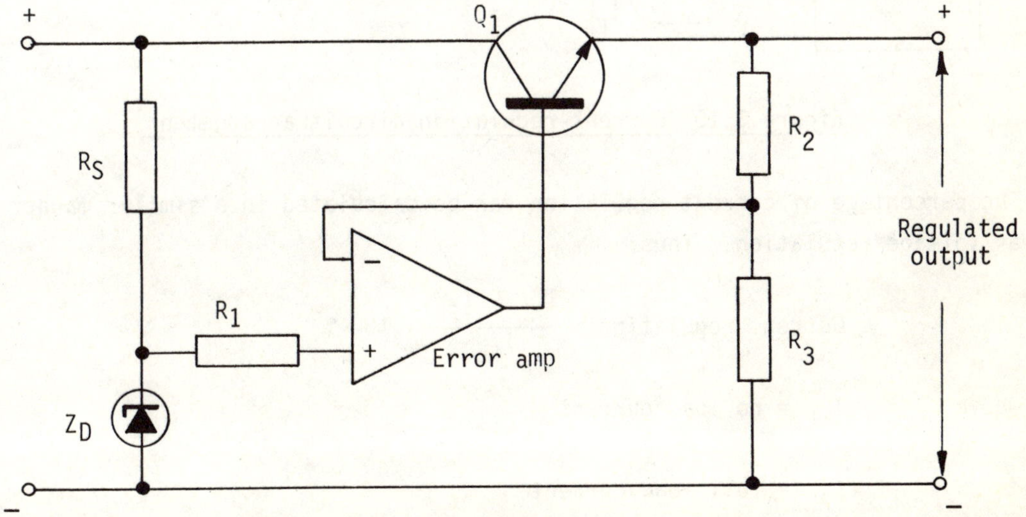

Figure 3.14 Regulator circuit using an op-amp as an error amplifier

3.7 CURRENT REGULATORS

A current regulator is designed to give a constant value of current through a load which has a varying load voltage. Ideally, the current with no load and the current with full load would be equal, but in practice this does not occur.

Remember, current will take the least resistive path around a circuit, and therefore, a no-load current is through a short circuit, and a load current is through the load. Figure 3.15 shows the circuit arrangement for these two conditions in order to determine the level of regulation.

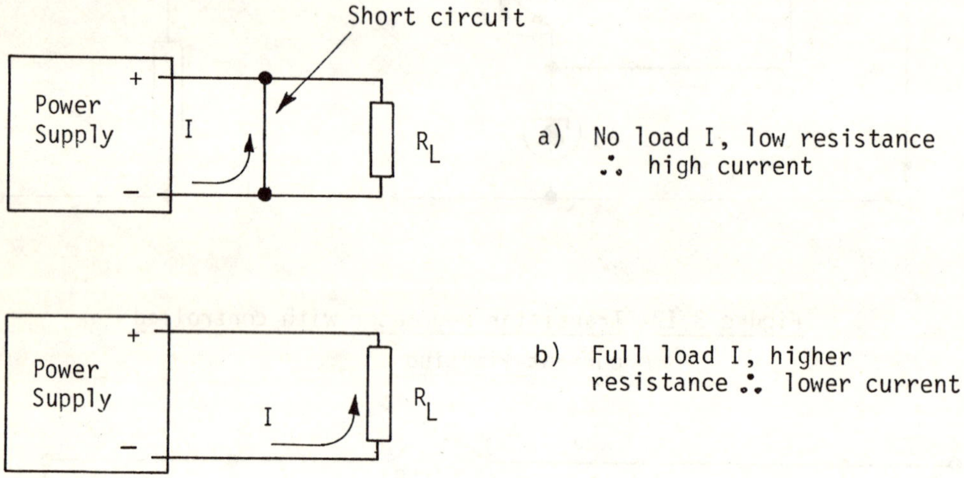

Figure 3.15 Current-regulation circuit arrangement

The percentage of current regulation can be calculated in a similar manner as voltage regulation. Thus:

$$\text{Current regulation} = \frac{I_{NL} - I_{FL}}{I_{FL}} \times 100\ \%$$

where I_{NL} = no load current

I_{FL} = full load current

Figure 3.16 shows the circuit of a constant-current regulator. A zener diode is connected across the series combination of R_1 and the base-emitter

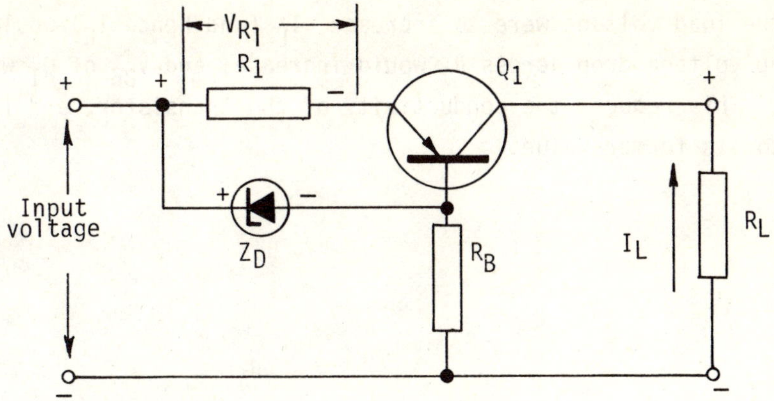

Figure 3.16 A series current regulator

junction. The emitter current (I_E) produces a voltage drop across R_1. The emitter current is limited to a value which causes a potential difference across R_1, which, when added to V_{BE} of Q_1, equals the breakdown voltage of the zener diode. Neglecting the 600 mV V_{BE}:

$$I_E = \frac{V_Z}{R_1}$$

R_1 can be made variable to allow adjustment of the current level.

The load current I_L equals the collector current I_C, which in turn equals the emitter current I_E minus the base current I_B. Hence:

$$I_L = I_C = I_E - I_B$$

and

$$I_E = I_C + I_B$$

Therefore, a decrease in I_L (due to a drop in load voltage) would cause an equal decrease in I_C. This would be reflected in the emitter current, and I_E would decrease. Any decrease in I_E reduces the voltage drop across $R_1 (V_{R1} = I_E \times R_1)$, and this affects the bias on Q_1:

$$V_{BE} = V_Z \text{ (fixed)} - V_{R1} \text{ (Note: observe the circuit polarities)}$$

Therefore, a decrease in V_{R1} would increase V_{BE} of Q_1, and the transistor will conduct more, maintaining the level of I_L fixed.

If the load voltage were to increase, I_L (and hence I_E) would increase also. The voltage drop across R_1 would increase, and V_{BE} of Q_1 would decrease. This reduces the conductivity of the transistor, and I_L would be reduced to its former value.

3.8 SELF-EVALUATION QUESTIONS (ANSWERS IN APPENDIX A)

1. Briefly explain why voltage regulation is required in a circuit.

2. A 18 V zener diode requires 15 mA to hold breakdown, and a circuit current of 85 mA ± 5 mA is required. If the input voltage is 25 V, calculate the value of series resistor required.

3. What is the minimum power rating of the zener diode used in Question 2?

4. Why are transistor shunt regulators only used in low-current applications?

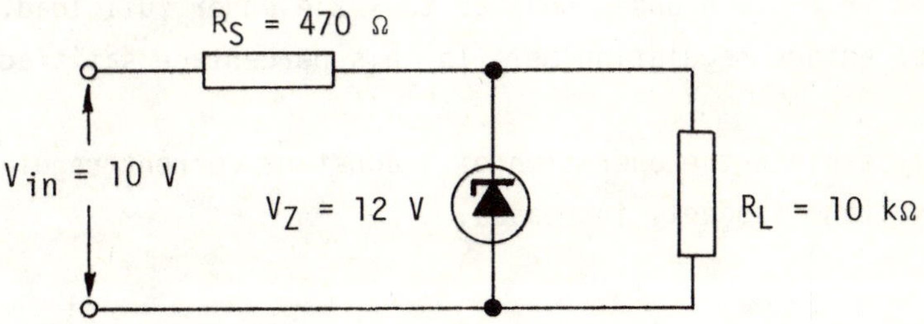

Figure 3.17

5. What is the output voltage of the circuit in Figure 3.17?

6. A series transistor regulator circuit (such as the one in Figure 3.7) has an input voltage of 15 V. The zener diode voltage is 12.5 V, and the zener current is 100 mA. If the load current is 1 amp, calculate:

   (a) The circuit efficiency.

   (b) The zener power rating.

   (c) The power dissipated by the transistor.

7. Briefly explain the circuit operation of a series transistor regulator when:

   (a) The load resistance decreases.

   (b) The input voltage increases.

8. How can a series regulator be protected against overload currents?

9. What is the difference between an 'open-loop' and a 'closed-loop' regulator circuit?

10. What is the purpose of a constant current regulator? If the current changed from 1.5 A under no load to 1.2 A under full load, what would the percentage regulation be? Is this percentage satisfactory?

11. Briefly explain the operation of a constant current regulator when the load voltage suddenly increases.

4 Integrated Voltage Regulators

4.1 INTRODUCTION

The trend in voltage regulation is towards localized regulation with integrated circuit (I.C.) regulators. They have the following advantages:

(1) Small size

(2) Low cost

(3) Low current

(4) Fixed voltage

I.C. regulators require minimal or no heat sinking, and few, if any, external components.

In the past, high-power regulators supplied all areas of an electronic system. These were large, costly and inefficient. Line impedance caused voltage drops which varied throughout the system. Unwanted coupling existed between critical parts of the system, and the excessive decoupling which this necessitated caused degraded local regulation. Now, recently developed three-terminal devices supplying 1 to 5 amperes can be placed in various areas of a system to satisfy the local area power requirements. These regulators are available in a variety of positive and negative voltages at various current ratings. Packages can be as small as the TO-92 plastic small-signal transistor or as large as the TO-3 power transistor. With this variety to choose from, it is now possible to select a regulator that is appropriate for each application, which will reduce cost significantly over other more established approaches.

Note: data, operation, and application notes in this chapter are reproduced with the kind permission of National Semiconductor from their 1978 Voltage Regulator Handbook. In order to keep information in this chapter consistent with the National Semiconductor data sheets, conventional current will be used. Later chapters will return to the use of electron flow.

4.2 PACKAGES

Available regulator packages are shown in Figure 4.1. The current capability of a regulator can usually be determined by the type of package

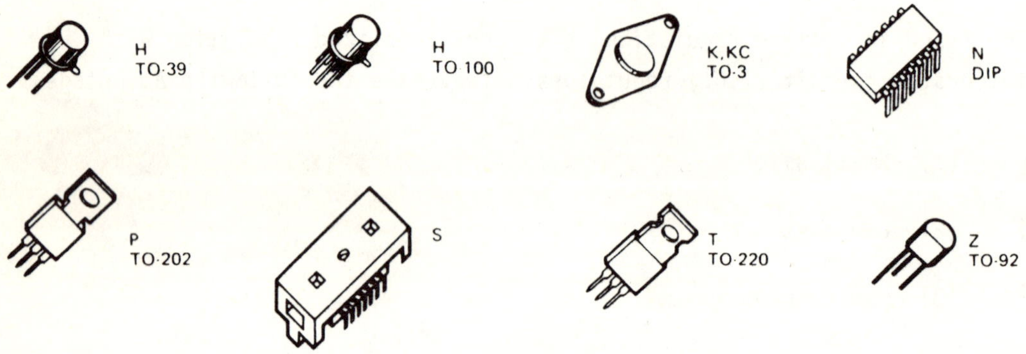

Figure 4.1 Available regulator packages

it is presented in. Some examples are shown in Table 4.1, but the final

Current	Package
100 mA	TO-92, TO-39
250 mA	TO-39, TO-202
500 mA	TO-39, TO-202
1.5 A	TO-220, TO-3
3.0 A	TO-3
5.0 A	TO-3

Table 4.1

selection of voltage and current capabilities should be made by referring to the manufacturer's selection chart or table. A typical chart presentation, listing data about three-terminal regulators is shown in Figure 4.2.

Output Current		Positive Output Voltage		Negative Output Voltage	
		Fixed Output Voltage	Adjustable Output Voltage	Fixed Output Voltage	Adjustable Output Voltage
5 amperes	Device Output Voltage Package		LM338 +1.2 V to +33 V TO-3		
3 amperes	Device Output Voltage Package	LM323 +5.0 V TO-3	LM350 +1.2 V to +33 V TO-3	LM345 -5.0 V, -5.2 V TO-3	
1.5 amperes	Device Output Voltage Package	LM340-XX, LM78XX + 5 V,+6 V + 8 V,+10 V, +12 V,+15 V, +18 V,+24 V TO-3,TO-220	LM317 +1.2 V to +37. V High Voltage (HV) +1.2 V to +57 V TO-3,TO-220	LM320-XX,LM79XX -5.0 V, -5.2 V, -6.0 V, -8.0 V, -9.0 V, -12 V, -15 V, -18 V, -24 V TO-3,TO-220	LM337 -1.2 V to -37 V High Voltage (HV) -1.2 V to -47 V TO-3,TO-220
500 milli-amperes	Device Output Voltage Package	LM341-XX, LM78MXX +5 V,+6 V, +8 V,+10 V, +12 V,+15 V, +18 V,+24 V. TO-202	LM317M +1.2 V to +37 V TO-202,TO-39	LM320M,LM79MXX -5.0 V, -5.2 V, -6.0 V, -8.0 V, -9.0 V, -12 V, -15 V, -18 V, -24 V. TO-202,TO-39	LM337M -1.2 V to -37 V TO-202,TO-39
250 milli-amperes	Device Output Voltage Package	LM342-XX +5 V,+6 V, +8 V,+10 V, +12 V,+15 V, +18 V,+24 V TO-202		LM320ML -5.0 V, -6.0 V, -8.0 V, -10 V, -12 V, -15V, -18 V, -24 V. TO-202	
100 milli-amperes	Device Output Voltage Package	LM340LA-XX, LM78L-XX +5 V,+6 V, +8 V,+10 V, +12 V,+15 V, +18 V,+24 V TO-39,TO-92		LM320L-XX, LM79L-XX, -5 V,-6 V,-8 V, -9 V,-12 V,-15 V -18 V,-24 V. TO-92,TO-39	

Figure 4.2 Three-terminal regulator selection guide

Once the voltage and current capability of the I.C. regulator is established, attention should be given to the terminal connections. *It is important to remember that similar packages do not have similar terminal connections.* Unless you are familiar with a particular regulator, always check the terminal connections by referring to the manufacturer's data.

4.3 CONNECTION DIAGRAMS AND CODING

The type of package, and hence the terminal connection, is determined by the order number of the regulator. For example, the LM117/LM217/LM317 family is available in a 500 mA rating (LM317M), and a 1.5 A rating (LM317). This gives a choice of four packages as shown in Figure 4.3a.

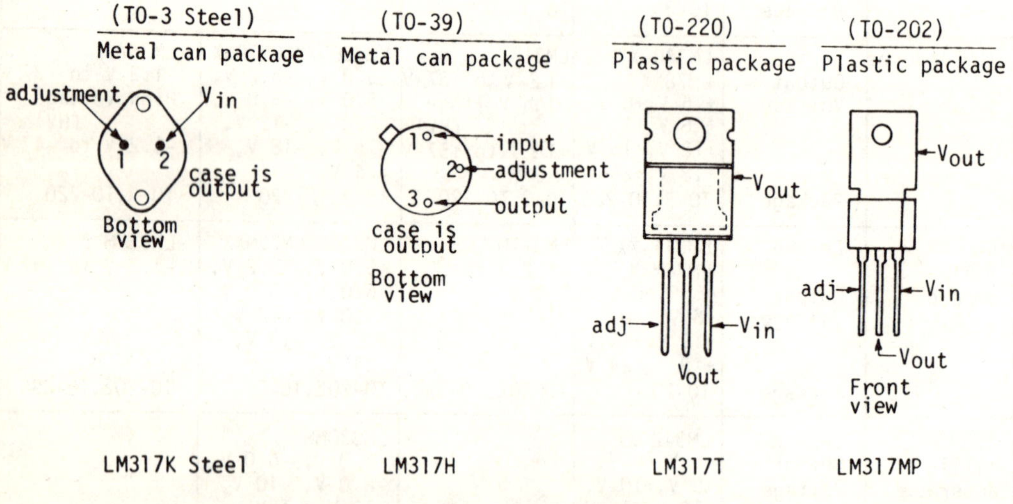

Figure 4.3a Four common regulator packages

The other package mentioned in the introduction to this chapter, TO-2, is shown in Figure 4.3b. A similar regulator, LM340T, is also available in a TO-220 package. Its specifications are: a fixed output 3-terminal regulator, with an output voltage range of +5 V to +24 V at 1.5 A. It looks identical to a LM317T, but the terminal connections are different. The terminal connections of the LM317T from left to right (see Figure 4.3a) are: Adjustment, V_{out} and V_{in}. However, the LM340T connections from left to right are: V_{in}, Common, V_{out}.

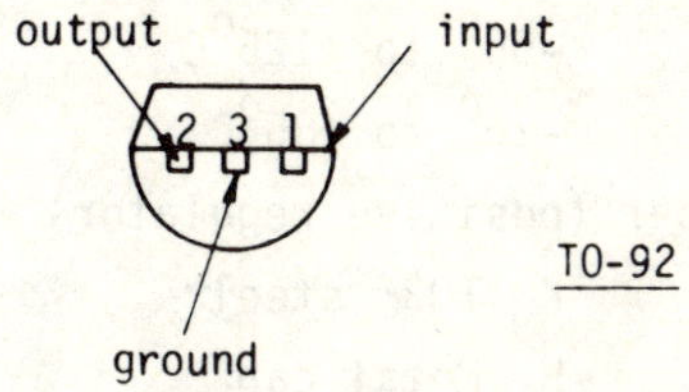

Figure 4.3b TO-92 plastic transistor pack

Type designation or coding. (The following information applies to National Semiconductor products only, and may be subject to variation.) The part number of the integrated voltage regulator is designed to give the following information:

1. Type of device (linear)

2. Temperature range of operation

3. Device code (fixed, variable, etc.)

4. Case style

5. Nominal output voltage

6. Tolerance (operating specification)

An example of this type of coding is given on the following page.

Part No.	LM	3	40	K	12	A
	1	2	3	4	5	6

1	LM	-	Linear Monolithic		
2	3	-	Temperature range:		0° to $+70^{o}$C
			(1)		-55° to $+125^{o}$C
			(2)		-25° to $+85^{o}$C
3	40	-	Basic part number (positive regulator)		
4	K	-	Case style	K	T03 steel
				H	Metal can
				N	D.I.P.
				P	Plastic (T202)
5	12	-	Output voltage :		+5 V to +24 V
6	A	-	High accuracy :		R = relaxed accuracy
					no letter = standard

Some regulators use a somewhat different designation, as shown below:

Part No.	LM	78L	05	AC	Z
	1	3	5	6	4

1	LM	-	Linear Monolithic
3	78L	-	Basic part number
5	05	-	Output voltage
6	AC	-	Improved accuracy - C - standard
4	Z	-	Case style Z TO-92 Plastic

There is no selection of operating temperatures, the only range being 0° to $+70^{o}$C.

4.4 HEATSINKING

Heatsinking is the technique used to transfer the heat created by power dissipation away from the I.C. If the regulator has a large difference between input and output voltage, then a fair amount of power will be dissipated by the I.C. itself. Assume that a 12 volt regulator handling a constant load current of 1.5 amperes has an input voltage of 30 volts. The power dissipated would be:

$$P = V \times I$$
$$= (V_{in} - V_{out}) \times \text{Current}$$
$$= (30 - 12) \times 1.5$$
$$= 27 \text{ watts}$$

This may well exceed the power dissipation rating of the regulator; if not, it will certainly cause it to run much hotter than is desirable and will cause automatic shut down. The excess heat can be reduced by attaching the regulator to a heatsink.

Heat can be transferred from the regulator package by three methods:

1. Conduction

2. Convection

3. Radiation

Conduction is the most effective method of moving heat from the I.C. junction to the case, and from the case to the heatsink. It can be aided by the use of a heat conductive paste placed between the regulator package and the heatsink.

Convection is an effective method of cooling both the regulator case and heatsink. It is dependent on the type of convecting agent (which in the case of a regulator is air), and it can be improved, if necessary, by increasing the air flow with the aid of a fan.

Radiation is the transfer of the heat from the cooling fins of the heatsink to the air. The choice and finish of material has a large bearing on the ability of the heatsink to get rid of the heat. A polished surface is not as efficient as an anodized or black-painted surface.

Care must be taken when connecting the regulator to the heatsink not to short the regulator output to ground. The TO-3 and TO-220 package have the case connected to the output terminal (see Figure 4.3a). They both require a mica insulating washer between the package and the heatsink when mounted.

There are many heatsinks available commercially, but sometimes it is more convenient and economical to mount the regulator to the chassis of the power supply, or to an aluminium extrusion. Whatever the method chosen, there are three basic rules for efficient heatsinking:

1. Mount the cooling fin vertically, where practical, for best convective heat flow.

2. Anodize or paint the fin surface area for better radiation.

3. Use 2 mm or thicker fins to provide low thermal resistance and hence better heat transfer from the regulator case.

4.5 BASIC THREE-TERMINAL REGULATOR

A standard three-terminal regulator connection is shown in Figure 4.4. All devices have short circuit protection, automatic thermal shut-down, on-chip

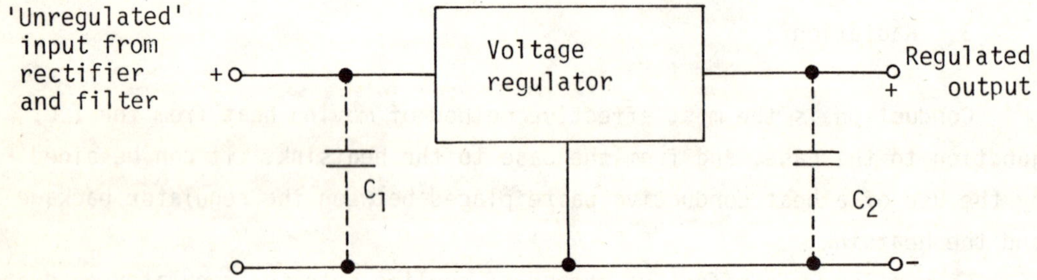

Figure 4.4 Normal three-terminal regulator connection

pass transistors and internal references. No external components are required. If the regulator is located more than 5 cm from the supply filter capacitor, a supply bypass capacitor (C_1) is required to maintain stability.

Capacitor C_1 (shown within dotted lines in Figure 4.4) should be a 100-200 nanofarad ceramic disc, or a 2.0 microfarad solid tantalum. It prevents any pickup on the long input lead from influencing the regulator's operation.

Capacitor C_2 is not absolutely necessary for stable operation, but a 1.0 microfarad tantalum capacitor gives the regulator both improved transient response and noise rejection. In some circumstances a 100 nF capacitor may be required across C_2 to minimize high frequency noise. (A 1.0 μF solid tantalum is equivalent to a 25 μF aluminium electrolytic at high frequencies.) With the basic regulator circuit connected as shown in Figure 4.4, a reliable, low-noise, well-regulated output voltage is obtained, provided that three other important conditions are met.

Firstly, the maximum power dissipation for the regulator must not be exceeded. Heat dissipated is wasted power, and a *sensible input-output voltage differential* is desirable. If the regulator is operated at too high a temperature, it will automatically shut down.

Secondly, the regulator must be operated above its 'drop-out' voltage. The drop-out voltage is the input-output voltage differential at which the circuit stops regulating. It is dependent on load current, and the regulator's temperature. A typical value for drop-out voltage is 1.5 V to 2.0 V.

From these two conditions it can be seen that there is an optimum range for the input and output voltage difference. The difference must not be too great, otherwise excessive temperature will shut the regulator down, and it must not be too small, otherwise it will fall below the drop-out voltage and the circuit will cease to regulate.

Thirdly, the amplitude of the ripple on the rectified, but unregulated input voltage must be considered. It must not fall into the 'drop-out' region on the negative-going half cycle or the regulator will falter in operation. Figure 4.5a shows a complete power supply circuit using a regulator.

The choice of capacitor for C_1 and C_2 has already been discussed; we now turn to C_F. The filter capacitor C_F must be large for two reasons:

1. To make the discharge time constant long (refer to Chapter 2, Section 2.5).

2. To keep the amplitude of the ripple low.

If a 12 volt regulator is operated at a minimum V_{in}-V_{out} differential of 1.5 V and the ripple amplitude is high (say 3 volts peak), then we have the situation shown in Figure 4.5b. The ripple intrudes below the regulator voltage and upsets the operation. To reduce the ripple amplitude, the value of C_F needs to be increased.

In Figure 4.5c, two things have happened: the V_{in}-V_{out} differential has been increased slightly (to 3 volts) and the ripple amplitude reduced. The regulator operates quite satisfactorily. The input-output voltage

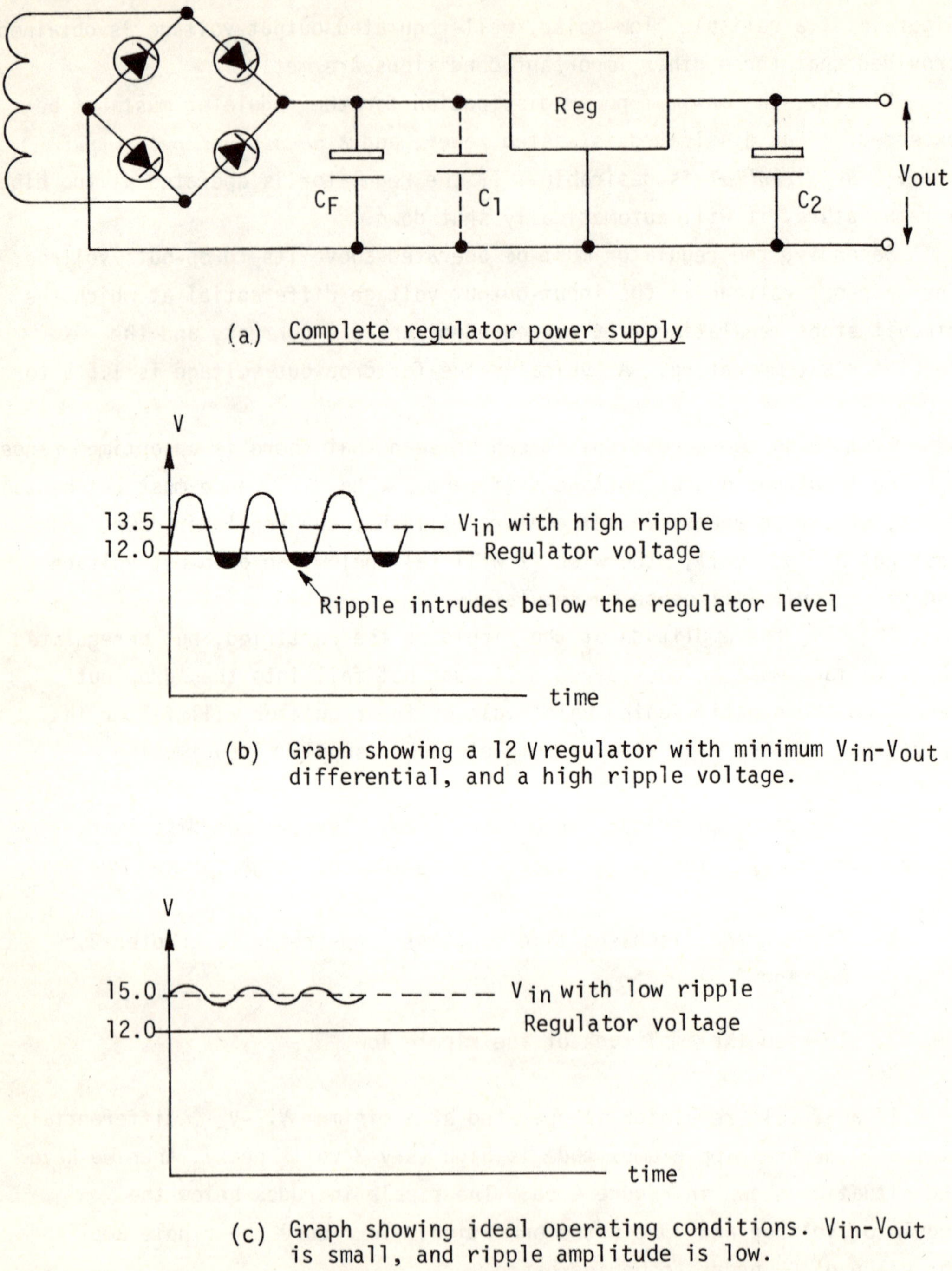

Figure 4.5 Regulator operating conditions

differential is sensible, and only a slight amount of power is dissipated. A typical value for C_F is 1000 μF at a voltage rating in excess of the rectifier output voltage.

Choke-input filters or π-section filters are rarely used for two reasons:

1. It is desirable to eliminate the weight and cost of a choke.

2. Regulator I.C.'s have a high rejection to ripple, and compensate for the shortcomings of capacitor input systems.

4.5.1 REGULATOR OPERATION

The circuit functions shown in Figure 4.6 are common to all three-terminal regulators. The reference voltage is a temperature-stabilized voltage, developed from a zener diode or a special 'band gap' circuit. The error amplifier compares the reference voltage with a portion of the output voltage. Resistors R_1 and R_2 act as a voltage divider, and the fraction of the output voltage, that is fed to the error amplifier, is determined by the feedback ratio of $R_2/(R_1 + R_2)$. The error amplifier output controls the base drive of the series pass transistor (Q_1) to provide regulation.

All the regulator protection circuits (current, safe area and thermal shut down) limit or turn off the base drive of the series pass transistor when activated. This either reduces the available output current or turns the series transistor off completely.

4.5.2 OPERATION OF THE INTERNAL PROTECTION CIRCUITS

Current Limiting. When the difference between V_{in} and V_{out} is less than the 6 V breakdown of the zener diode, there is no current in R_3 and only a small Q_2 base current in R_4. The base-emitter voltage of the current limit transistor Q_2 equals the voltage developed across the current sensing resistor R_{CL} (Figure 4.7).

As the regulator's output current increases, the voltage across R_{CL}, and hence the V_{BE} of transistor Q_2, also increase. Eventually, Q_2 turns on. This prevents additional base drive from reaching the series pass transistor (Q_1) and therefore limits the output current.

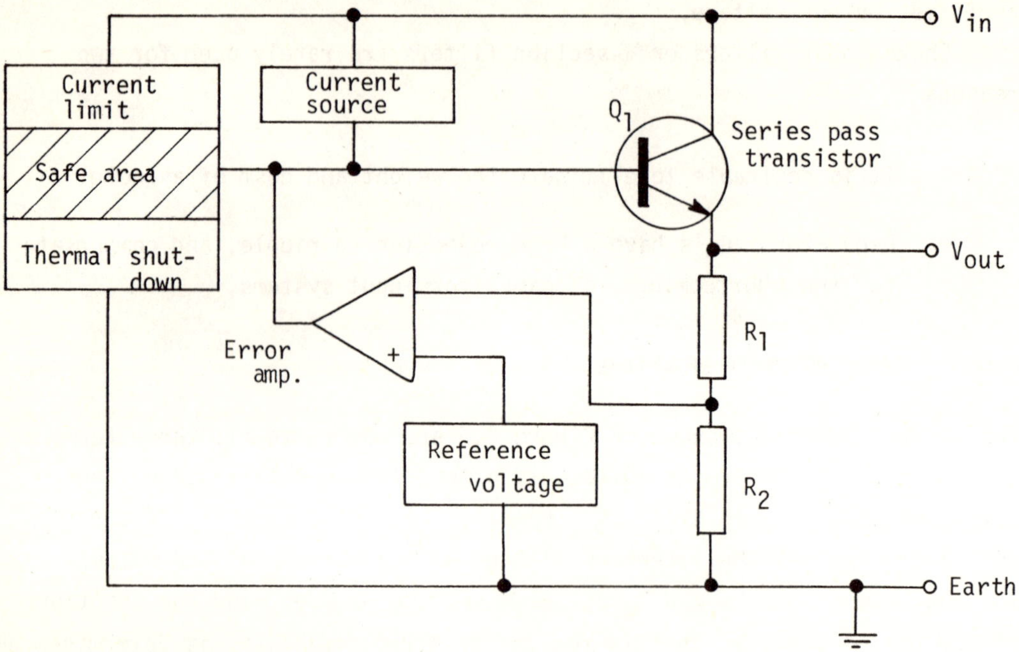

Figure 4.6 Basic regulator arrangement

Thermal Shutdown. Transistor Q_3 in Figure 4.7 is the 'thermal shutdown' transistor. It is physically located next to the series pass transistor Q_1, which is the major heat source in the regulator package.

The base of Q_3 is held just below its turn-on voltage at room temperature (approximately 400 mV). As the temperature of the package increases, the voltage required to turn Q_3 on will decrease to 400 mV. At this point, the transistor Q_3 turns on; this removes the base drive from the series pass transistor and turns the output off.

Regulators have thermal shutdown temperatures ranging from 150^oC to 190^oC. If the package body exceeds the temperature rating, the I.C. automatically turns off.

Safe Area Protection (Refer to Figure 4.7). When the difference between V_{in} and V_{out} is greater than the breakdown voltage of the zener diode, a current which is proportional to this voltage difference (V_{in}-V_{out}) passes through the zener diode, R_3 and R_4 to the output. This causes the

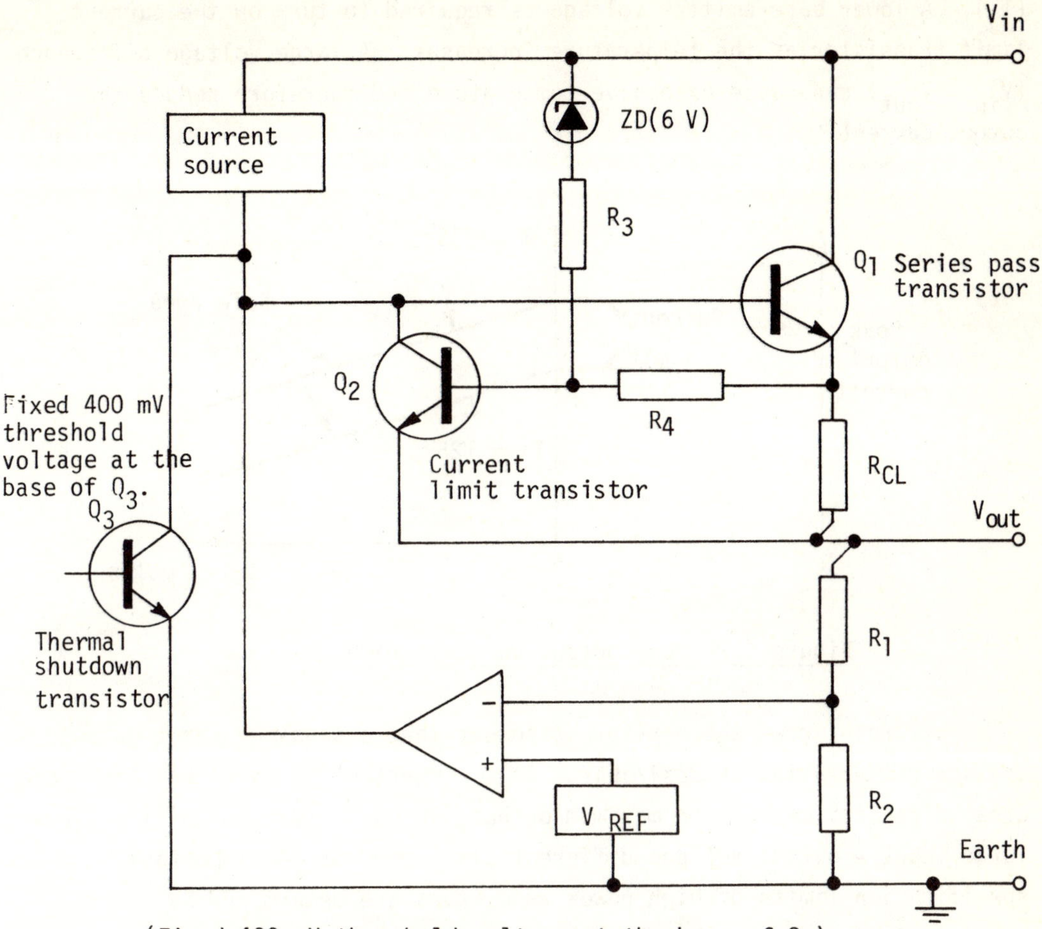

(Fixed 400 mV threshold voltage at the base of Q_3)

Figure 4.7 Basic regulator with protection circuits

base-emitter voltage of transistor Q_2 to be greater than the voltage drop across R_{CL}. Transistor Q_2, therefore, turns on at a lower output current through R_{CL}, and the current limit point of the regulator is reduced.

The rate of reduction of current limit, with increase in input-output voltage differential, is equal to:

$$\frac{\Delta I_{CL}}{\Delta (V_{in} - V_{out})} = \frac{R_4}{R_3 R_{CL}} \text{ amperes per volt}$$

where Δ means 'a change in'.

The curves shown in Figure 4.8 indicate the safe area characteristics. They show a reduction in output current with increased junction temperature

(T_j). A lower base-emitter voltage is required to turn on the current limit transistor as the temperature increases. A large voltage difference (V_{in} - V_{out}) can cause excessive temperature and therefore reduce the output current.

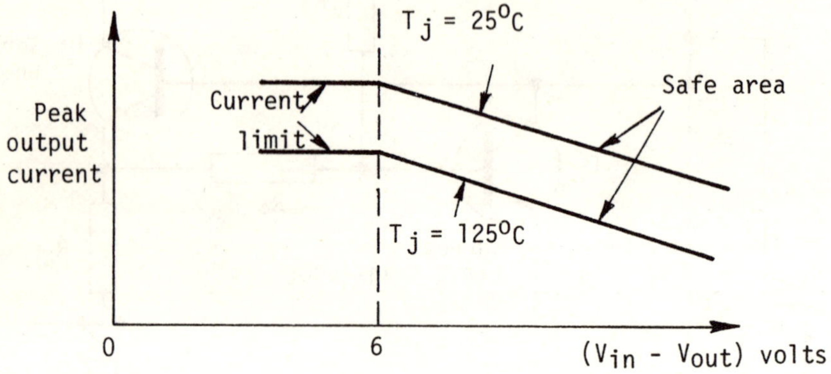

Figure 4.8 Peak output current graph

This reinforces the earlier statement that a sensible input-output voltage differential is desirable. It is important to note that the 'safe area' circuitry causes the maximum output current to drop significantly with large input - output voltage differentials. Many of the important specification limits of high power regulators are determined by thermal characteristics rather than the electrical characteristics.

4.6 APPLICATIONS

Voltage regulator use can be extended beyond that of the simple three-terminal fixed voltage device just discussed, and some useful practical circuits will now be described. Equations relative to the circuit operation are included which will enable the concepts to be applied to the entire regulator family. But first, three handling precautions are given.

4.6.1 PRECAUTIONS WHEN USING REGULATORS

A range of voltage regulators are designed with thermal, short circuit, and safe area protection. However, with any integrated circuit regulator it is necessary to take precautions to ensure that the regulator is not inadvertently damaged.

1. *Shorting the regulator input.* When using large capacitors at the output of regulators that have an output voltage greater than 6 volts, a precaution is required. A protection diode D (shown in Figure 4.9) connected between input and output may be required if the input is shorted to ground.

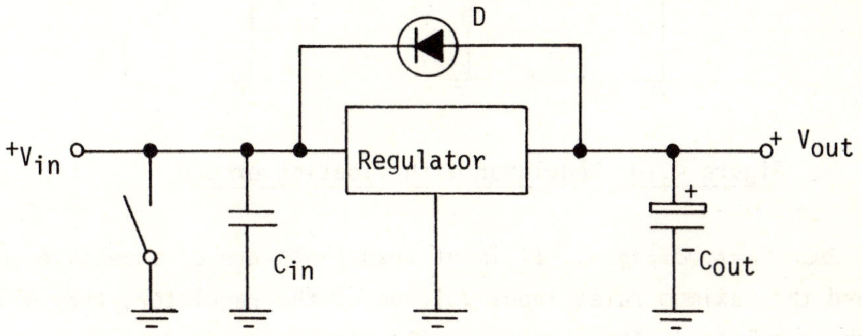

Figure 4.9 Shorted input

An input short will pull the input to ground potential, while the output remains close to V_{out} because of the charge stored in the output capacitor. The capacitor will try to discharge through the low-impedance junction inside the regulator. If the energy is large enough, both the junction and the regulator will be destroyed. However, a fast diode (D in Figure 4.9) would shunt the capacitor's discharge current around the regulator and protect it.

If the output voltage of the regulator is raised above the input voltage, then the internal current paths of the regulator will be damaged in a manner similar to that just described. The diode shown would also protect the I.C. by becoming forward biased and shunting the regulator's internal circuitry.

2. *Regulator floating ground.* If the ground pin alone becomes disconnected (as shown in Figure 4.10), during servicing or construction, the output approaches the unregulated input voltage level. This could cause damage to any circuit connected to V_{out}. If the open ground connection is remade with the power 'ON', the regulator itself may be damaged. This fault is most likely to occur when plugging regulators (or modules with on-card

regulators) into powered-up sockets. Power should be turned 'OFF' prior to removal or insertion of regulator I.C.s.

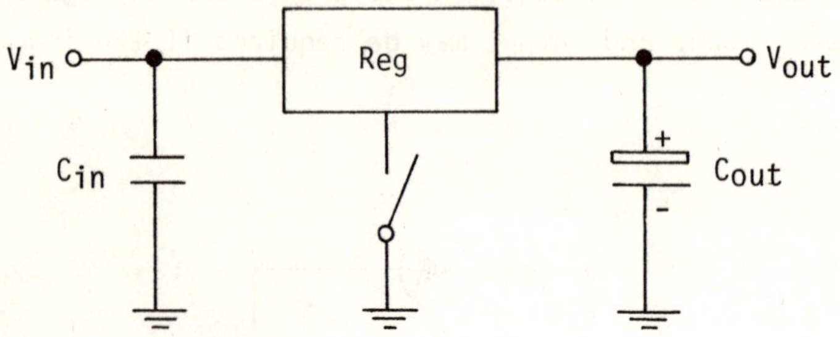

Figure 4.10 Regulator with floating ground

3. *Transient voltages.* If input transients are of excessive amplitude and exceed the maximum rated input voltage of the regulator, they will damage the regulator. The regulator will also be damaged if the transients reach more than 800 mV below ground, i.e. negative input.

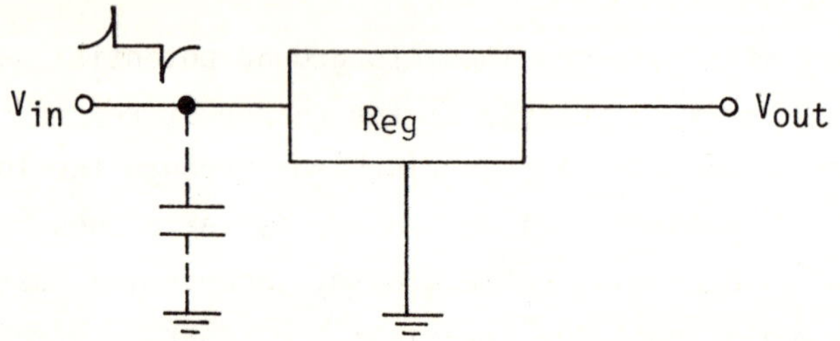

Figure 4.11 Transients

The solution to this problem is to use one or a combination of the following components: a large input capacitor (see Figure 4.11), a series breakdown diode, a choke, or a transient suppressor.

4.6.2 POSITIVE REGULATOR CIRCUITS

Note: Conventional current is indicated in all regulator circuits.

Figure 4.12 is the basic three-terminal voltage regulator as described in Section 4.5. As discussed then the choice of C_1 and C_2 depends on the circuit's environment and the constructor's circuit layout.

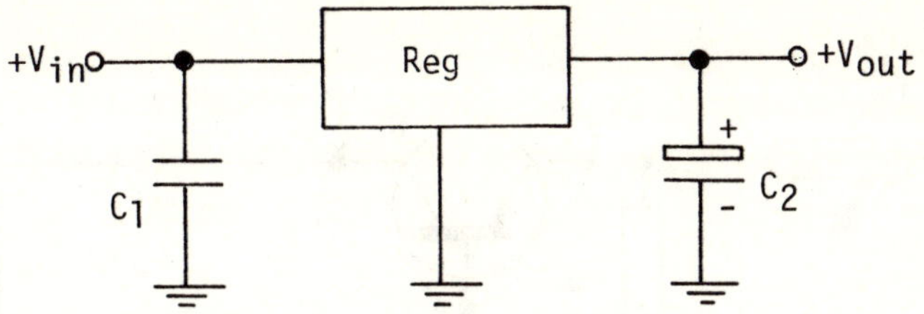

Figure 4.12 Basic regulator as described in Section 4.5

Positive regulator as a current source. Figure 4.13, an I.C. regulator used as a current source. A constant output current I_L is delivered to a variable load.

$$I_L = \frac{V_{Reg}}{R} + I_Q$$

where I_Q is the quiescent current, and V_{Reg} is the regulator voltage rating. The impedance to be used as a load can be calculated from:

$$\text{Load } Z \leq \frac{V_{in} - (V_{Reg} + V_{dropout})}{I_L}$$

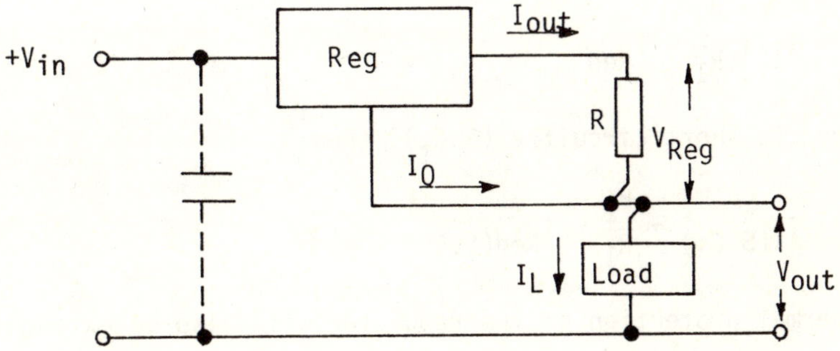

Figure 4.13 Current source

High current regulator. The current boost circuit shown in Figure 4.14 takes advantage of the internal current-limiting characteristics of the regulator. It provides short circuit current protection for the series pass transistor as well.

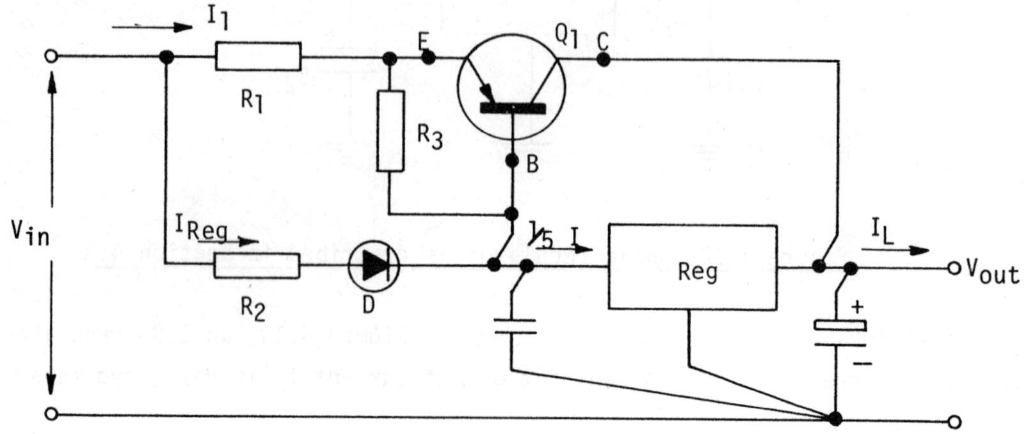

Figure 4.14 High current regulator with output short circuit limit

The regulator I.C. handles 1/5th of the output current and provides voltage regulation, while Q_1 handles the remaining higher output current. The ratio of current sharing between the regulator and the transistor is set by the ratio of R_1/R_2, provided that $V_D = V_{BE}$ of Q_1, where V_D is the voltage drop across the diode.

$$I_1 = \frac{R_2}{R_1} \times I_{Reg}$$

If the output is short circuited (S.C.), then:

$$I_{1(S.C.)} = \frac{R_2}{R_1} \times I_{Reg(S.C.)}$$

The thermal protection of the regulator will also be extended to the transistor if two conditions are met:

1. The regulator and Q_1 have the same thermal resistance: i.e. junction-to-ambient temperatures the same.

2. The pass transistor heatsink has R_2/R_1 times the thermal capacity of the regulator heatsink.

Component suggestions are listed below.

Q_1	D	I_1	I_{Reg}	R_2/R_1	R_3
2N4398	IN4719	≥ 3 A	1 A	≥ 3	5-10 Ω
NSD32	IN4719	2 A	1 A	2	5-10 Ω
NSDU51A	IN4003	1 A	500 mA	2	5-10 Ω

Using the circuit configuration in Figure 4.14, there is an increase in the minimum input - output voltage differential of the regulator. The voltage drop across the diode and R_1 must be taken into consideration. The minimum input - output voltage differential is increased by one diode voltage drop (i.e. V_{BE} of Q_1), plus the voltage developed across resistor R_1 (I_1R_1).

The series pass transistor (Q_1) for high current application is an expensive PNP type. A low-priced PNP/NPN combination can be used in its place. Figure 4.15 shows the configuration and the circuit connection points are E.B. and C.

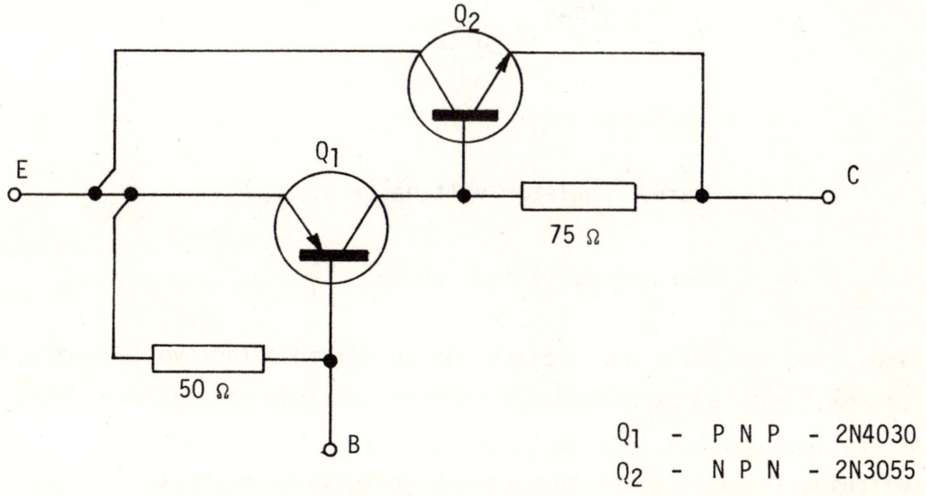

Figure 4.15 Replacement circuit configuration for high-current application

Adjustable output voltage. In Figure 4.16 a fraction of the regulator current (V_{Reg}/R_1) is used to raise the ground pin of the regulator above earth. This is achieved by a voltage drop across $R_2\left[V_{Reg}/(R_1 \times R_2)\right]$,

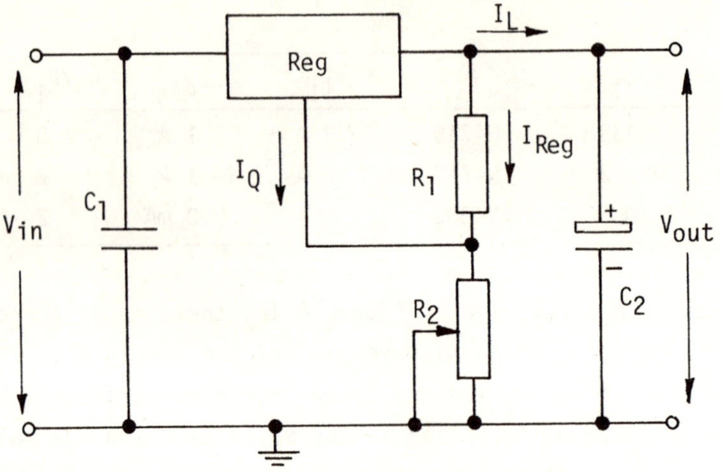

Figure 4.16 Adjustable V_{out}

and it provides an adjustable output voltage. However, it will not turn down to zero volts.

$$V_{out} = V_{Reg} + \left(\frac{V_{Reg}}{R_1} + I_Q\right)R_2$$

where

V_{Reg} = rated regulator voltage (e.g. 5 V)

I_Q = quiescent operating current

These are but a few applications of the fixed-output voltage regulator, and further ideas can be obtained from data published by National Semiconductor. One circuit that is becoming very popular in modern equipment is the switching regulator, and this will be discussed in Chapter 7.

4.6.3 NEGATIVE REGULATOR

With the advent of operational amplifiers and microprocessors requiring dual-polarity power supplies, a range of fixed negative voltage regulators is also available. Figure 4.17 shows a three-terminal negative regulator with basic decoupling components and input short-circuit protection.

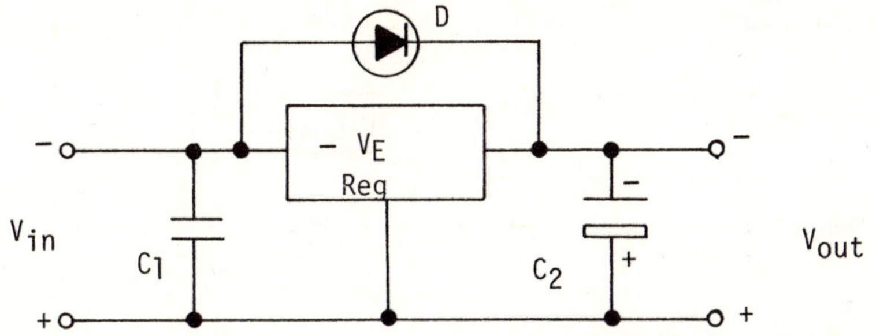

Figure 4.17 Basic negative regulator

Capacitors C_1 and C_2 and diode D perform the same function as they do in a positive regulator circuit. The only variation is the reverse connection of the diode and the polarity of the bypass capacitors. The basic positive regulator circuits shown in Figures 4.15, 4.16, and 4.17 can be duplicated with a negative regulator as shown in Figures 4.18, 4.19 and 4.20. However, the output voltage is negative, and conventional current is in the opposite direction.

4.7 DUAL POWER SUPPLY

A positive regulator, such as LM340, can be connected with a negative regulator, LM320, to provide a non-tracking dual power supply. Each regulator exhibits the specifications and limitations of the individual device. The circuit connection is shown in Figure 4.21. Diodes D_1 and D_2 are protection diodes to allow the regulator to work with a common load, and they should be rated at the short circuit current rating of the regulator.

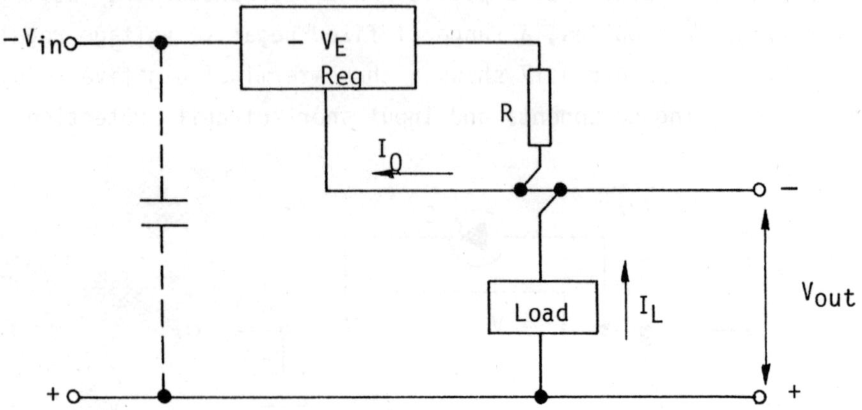

Figure 4.18 Current source

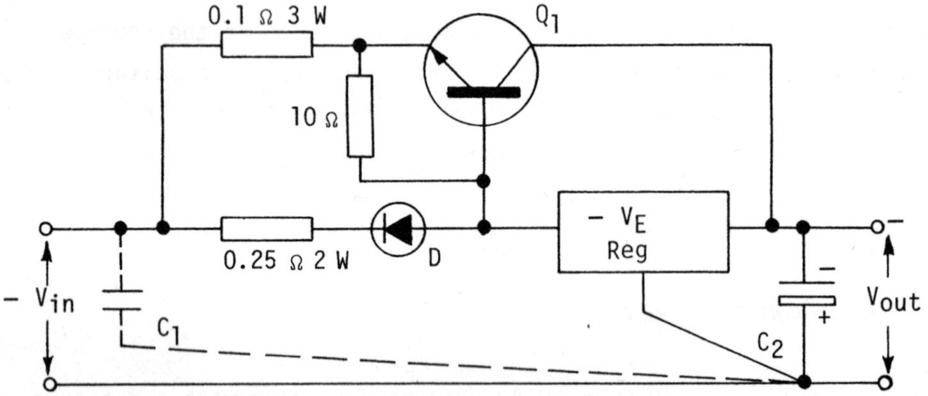

D = IN4720, Q = 2N3055 (I = 5 A), $-V_E$ Reg = LM320

Figure 4.19 High current regulator

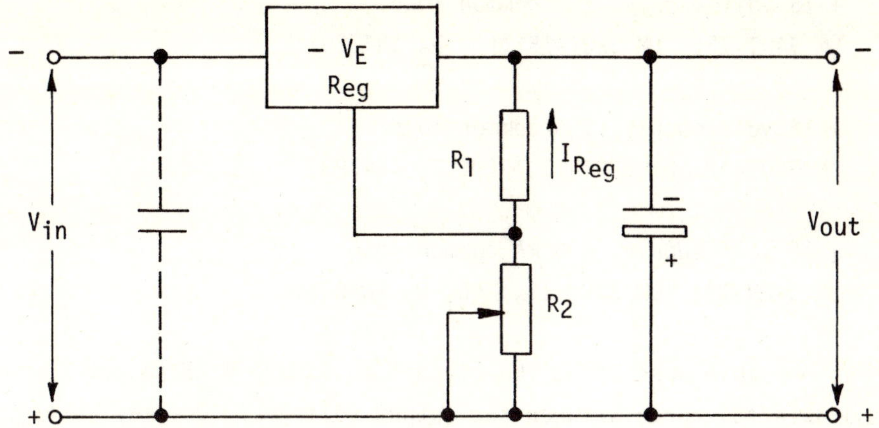

Figure 4.20 Adjustable output

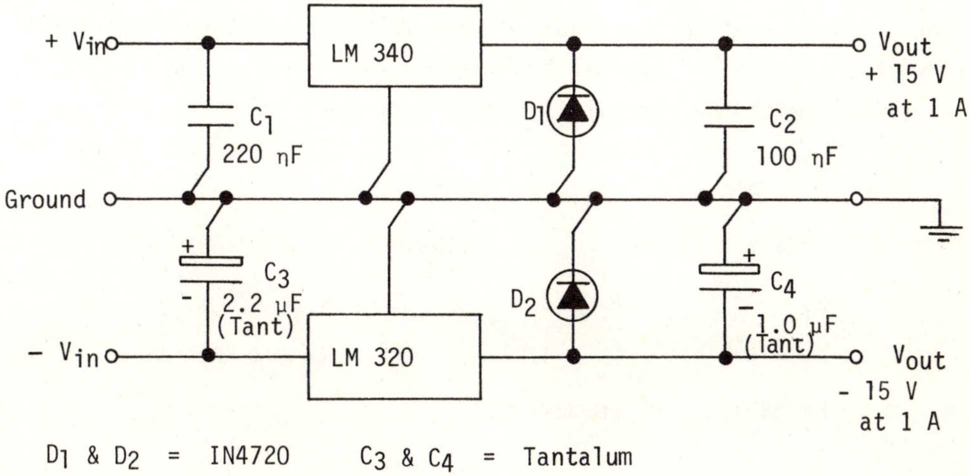

Figure 4.21 Basic dual supply

Typical combinations of positive and negative regulators are:

1. ± 15 volt supply, 1 A common load.
   LM 340T-15, LM 320T-15, (D_1 D_2 IN4720)

2. ± 12 volt supply, 1 A common load.
   LM 340T-12, LM 320T-12, (D_1 D_2 IN4720)

3. ± 15 volt supply, 200 mA common load
   LM 342H-15, LM 320H-15, (D_1 D_2 IN4001)

It is possible when using the circuit in Figure 4.21 to get a wider than desirable tolerance between the output voltages. One may be +15.5 V and the other -14.5 V. This can be overcome by modifying the basic circuit by providing a means of trimming the output voltages to a closer tolerance.

In the circuit in Figure 4.22, potentiometers R_4 and R_6 allow a trimming voltage to be developed across resistors R_1 and R_2. Resistors R_3 and R_5 help to linearize the adjustment and prevent shorting between the regulator ground pin and the opposite polarity output voltage. D_3 is a 'germanium' diode that may be required to enable the positive regulator to start with a high load. Typical component values for those who would like to experiment with the circuit in Figure 4.22 are:

C_1 = 220 nF	$D_1 \times D_2$ = IN4720	
C_2 = 2.2 μF	D_3 = IN91	
R_1 = 33 Ω	R_2 = 33 Ω	R_3 = 220 Ω
R_4 = 5 k Ω	R_5 = 470Ω	R_6 = 2 kΩ
LM 340T-5,	LM 320T-5.2	

Ideally, as with our basic transistor voltage regulator, we want a circuit that automatically adjusts to compensate for variations in the output condition. When using a dual supply we want the two voltages to lock together, and if there is any variation in one, a corresponding variation takes place in the other. The circuit that operates in this fashion is called a 'tracking dual supply'.

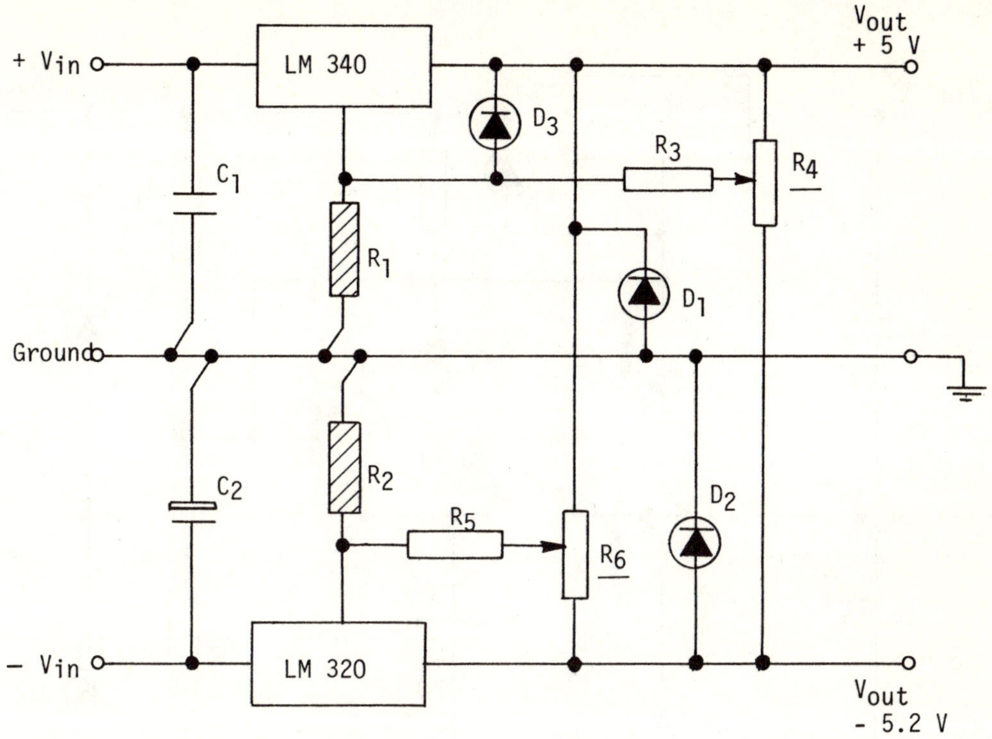

Figure 4.22 Dual supply with trimming potentiometers R_4 and R_6

4.7.1 TRACKING DUAL SUPPLY

The circuit of a tracking dual supply is shown in Figure 4.23. The positive regulator 'tracks' the negative regulator. Diodes D_1, D_2 and D_3 are the same as those used in Figure 4.22. R_1 and R_2 are identical and of close tolerance (1%) for close tracking.

Point V_A is at virtual ground potential under steady state operating conditions, and transistor Q_2 passes the quiescent current of the positive regulator. If the output voltage of the negative regulator falls, point V_A follows accordingly, forward biasing the collector-base junction of transistor Q_1. The emitter voltage (Point V_B) of Q_1 decreases. The forward bias on transistor Q_2 is reduced, and the collector voltage therefore increases. This increase in collector voltage is also an increase in the positive regulator output voltage. The voltage divider of R_1 and R_2 is across the positive/negative output, and therefore an increase in + V_{out} restores the voltage at Point V_A to its former level.

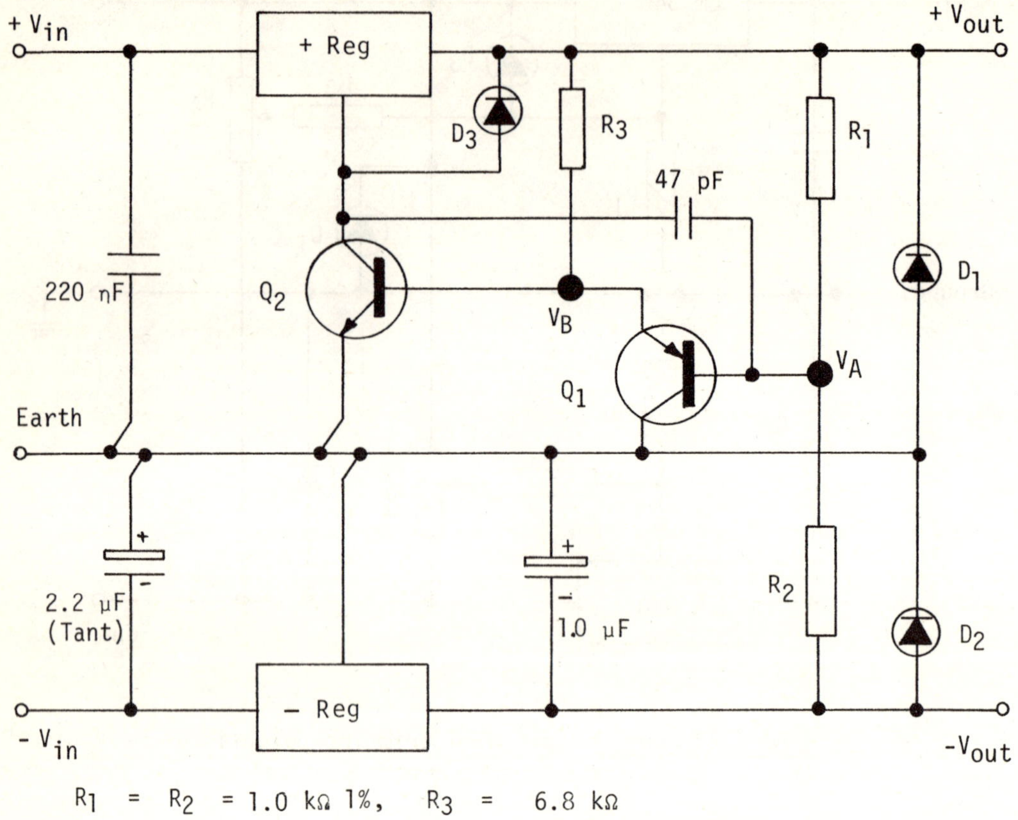

Figure 4.23 Tracking dual regulator

A typical combination for a ± 15 volt tracking dual supply is: LM340T-5, LM320T-15. The LM340T will track the LM320T within 100 millivolts if R_1 is matched to R_2 to within one percent.

Tracking regulators are also available fully encapsulated in a dual-in-line package (D.I.L.) compete with heatsink if required. The only connections required to the package, other than the plus and minus voltage inputs, outputs, and earth, are the completion of the output current paths for plus and minus current limiting.

4.8 ADJUSTABLE-POWER REGULATORS

A relatively new concept in three-terminal adjustable regulators is the power regulator. The LM317 is capable of supplying in excess of 1.5 A over a 1.2 V to 37 V output range, or up to 57 V in the high voltage mode. The LM350 is capable of supplying in excess of 3 A over a 1.2 V to 33 V output range. The LM338 is capable of supplying in excess of 5 A over a 1.2 V to 33 V output range.

These and other regulators, both fixed and adjustable, both positive and negative, are listed in the selection guide in Figure 4.2. They are easy to use and require only 2 external resistors to set the output voltage. Regulation is extremely good, and all include the standard current-limiting, safe-area, and thermal-limiting features of the basic fixed output regulator. At the time of writing a new device is being tested. It is a true power regulator, housed in a T03 case. It will make available an adjustable output voltage from 1.2 V to approximately 20 V, and perhaps even higher, at a current in excess of 10 A. It has similar features and specifications to the LM350 and only requires 2 external resistors to set the output voltage. The part number is LM396.

4.8.1 OPERATION

The adjustment of a three-terminal regulator is shown in Figure 4.24. An operational amplifier connected as a buffer drives a power Darlington. The circuitry is arranged so that all the quiescent current is delivered to the regulator output (rather than earth), eliminating the need for a separate ground terminal. A reference diode D_1 inserts 1.2 V between the non-inverting input of the op-amp and the adjustment terminal. Approximately 50 μA is required to bias the reference diode.

When operating, the output of the regulator is the voltage at the adjustment terminal plus 1.2 volts. If the adjustment terminal is grounded, the circuit behaves as a 1.2 V regulator.

To increase the output voltage, a divider R_1 and R_2 is connected from the output to ground as shown in Figure 4.25. The 1.2 V reference across R_1 forces about 5 mA of current flow. This 5 mA flows through R_2, increasing the voltage at the adjustment terminal and hence the output voltage:

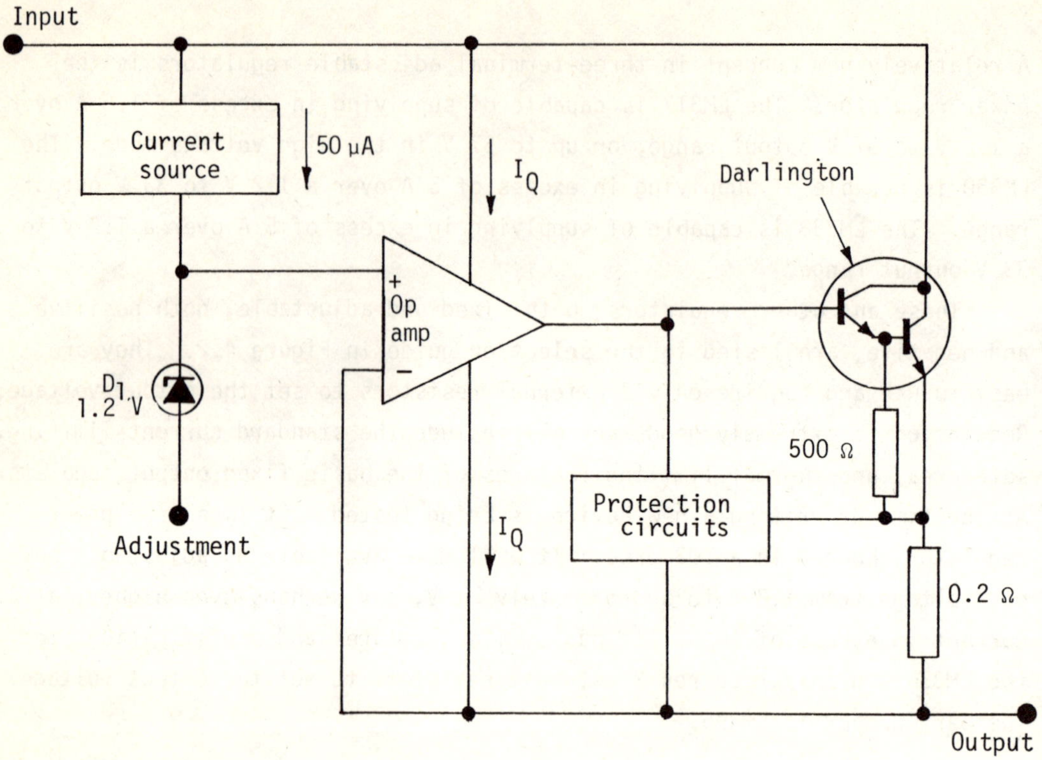

Figure 4.24 Functional diagram of an adjustable regulator (LM 317)

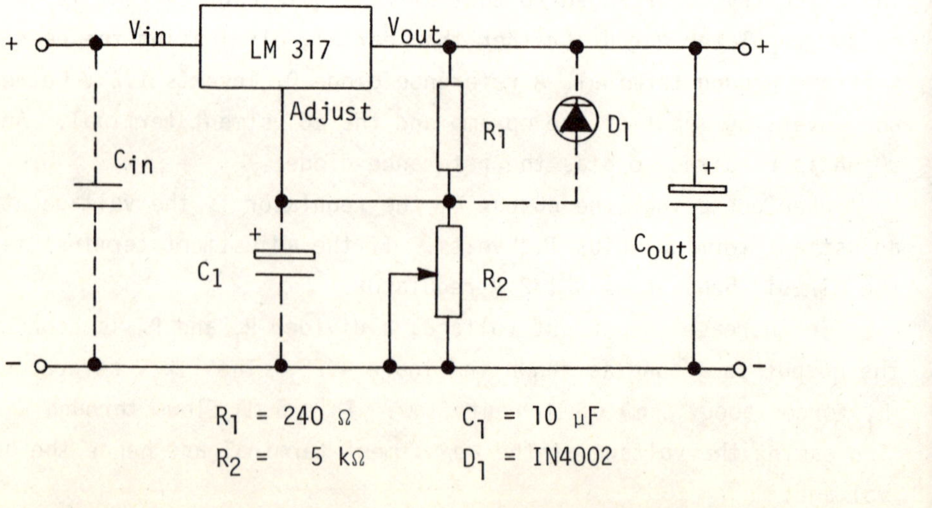

Figure 4.25 Voltage divider added to increase the output voltage

$$V_{out} = 1.2\ V \times (1 + \frac{R_2}{R_1}) + 50\ \mu A\ R_2$$

Capacitor C_1 can be included to improve ripple rejection, and diode D_1 will protect the regulator against short circuits at the input or output. The circuit shown in Figure 4.25 makes quite a suitable basic 1.5 A workshop power supply. The values of C_{in} and C_{out} were discussed in Section 4.5. The supply is adjustable from 1.2 V to 25 V, depending on the value of the input voltage, and the maximum $V_{in} - V_{out}$ is 40 V.

If a higher-current, higher-voltage supply is required, a LM350 or LM338 can be used. However, to ensure complete protection of the regulator, the precautions mentioned in Section 4.6.1 should be heeded. A further feature would be to make the regulator adjustable from zero volts. This can be achieved by returning the adjustment pin to a negative reference voltage. Figure 4.26 shows a typical circuit with full protection.

4.8.2 TYPICAL APPLICATIONS

Applications for a high-current adjustable voltage regulator include: temperature controller, high-stability regulator, high-current regulator, adjustable current regulator and many others. Circuits for these applications are shown in the *National Semiconductor Voltage Regulator Handbook (1978)*, however a couple of useful and interesting ones are included here for experimenters.

The output voltage for the circuit in Figure 4.27 can be determined from the formula:

$$V_{out} = 1.2 + (1 + \frac{R_2}{R_1}) + 50\ \mu A\ R_2$$

$$V_{out} = 1.2 + (\frac{1100}{240} + 1) + 50 \times 10^{-6} \times 1.1 \times 10^{3}$$

$$= 1.2 + 5.58 + 55 \times 10^{-3}$$

$$= 6.8\ V$$

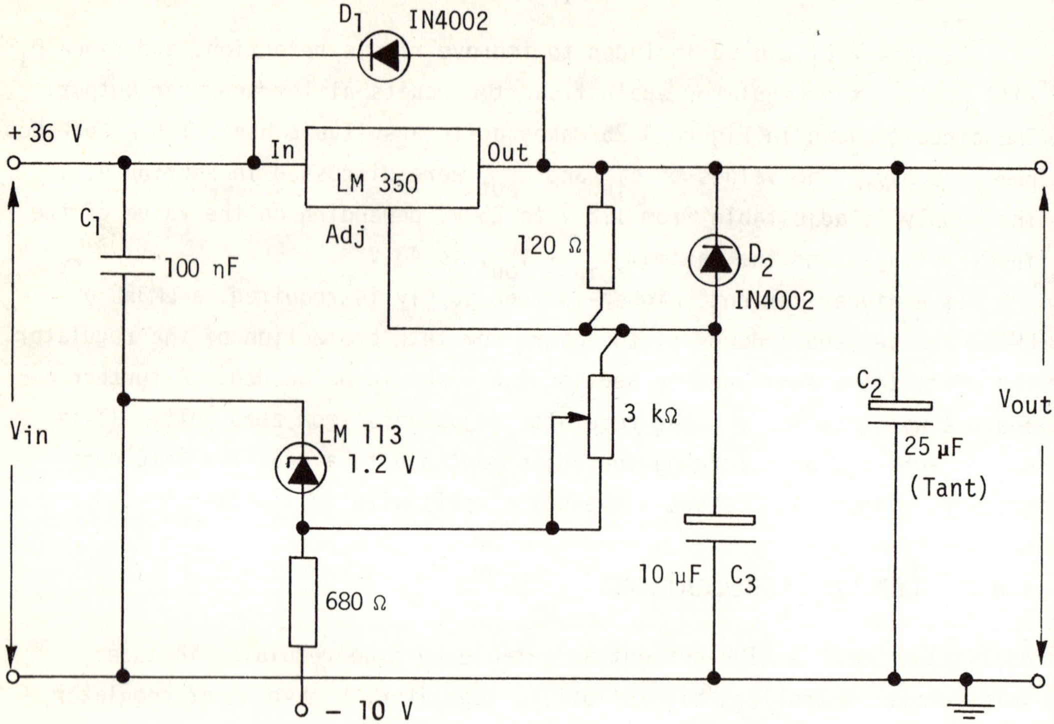

D_1 protects against C_2

D_2 protects against C_3

Figure 4.26 0 to 30 V regulator at 3 ampere with full protection

The 0.3 Ω resistor sets the maximum current of the circuit by controlling the bias on transistor Q_1.

$$V_{BE} = I \times R(0.3\ \Omega)$$

$$= 2 \times 0.3$$

$$= 0.6 \text{ volts}$$

$$= 600 \text{ mV}$$

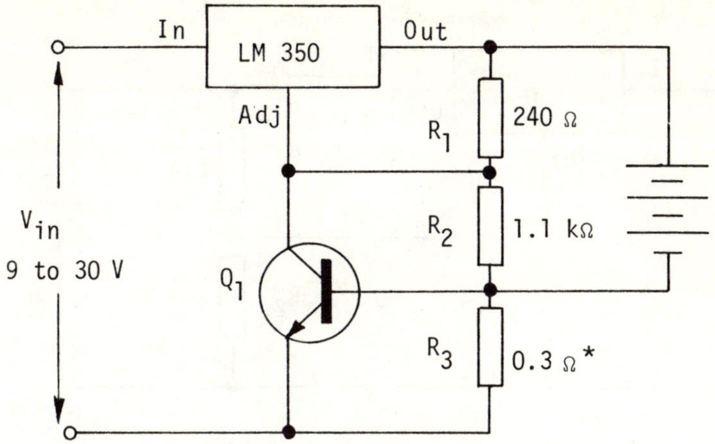

*Sets peak current (0.3 Ω for 2 A)

Figure 4.27 A 6 volt current-limited battery charger

Therefore the voltage across the battery is 6.2 volts.

If the current was to suddenly try to increase, the voltage developed across the 0.3 Ω resistor would increase. This would increase the forward bias (V_{BE}) on Q_1. Transistor Q_1 is in parallel with $R_2 + R_3$, and therefore, as its conduction increases, the total resistance of the parallel network decreases. The voltage developed across this network is fed to the adjustment pin of the regulator to set the output voltage, and therefore this also decreases. The output of the regulator is thereby reduced, protecting the battery. For example, if the resistance of the parallel network drops to 500 Ω, the output of the regulator decreases to approximately 4 volts.

4.9 BASIC POWER-SUPPLY CIRCUITS

The purpose of this section is to provide some practical circuit configurations, showing the complete circuit from input to output. The final choice of transformer, rectifier configuration and filter is up to the circuit builder, but the following circuits should provide some interesting possibilities.

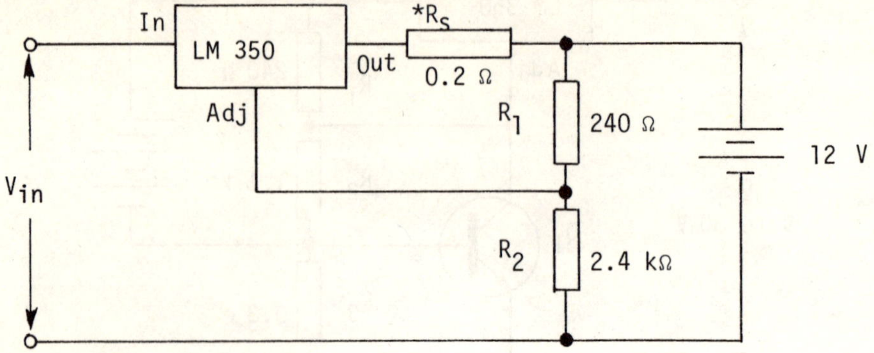

*R_s sets the output impedance of the charger, $Z_{out} = R_s(1 + R_2/R_1)$. This enables a lower charging rate for fuller batteries.

Figure 4.28 Simple 12 volt battery charger

The circuit in Figure 4.29 represents a full-wave centre-tapped rectifier using a capacitive filter. It is the most common selection for medium-power, regulated D.C. supplies.

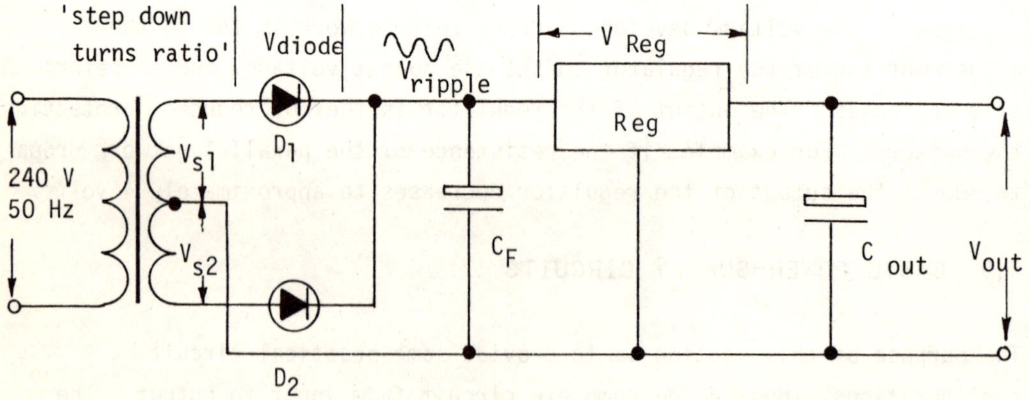

Figure 4.29 Full-wave centre-tap circuit

The full-wave bridge circuit, as shown earlier in Figure 4.5a, can also be used. This configuration is popular, especially given the availability of bridge rectifier packages.

A circuit not previously mentioned in this book is a dual full-wave circuit. It is called a 'dual complementary rectifier circuit', and it is a very efficient way of obtaining two identical outputs of reverse polarity, sharing a common ground. This circuit configuration is shown in Figure 4.30. It is also known as a 'centre-tapped bridge rectifier'.

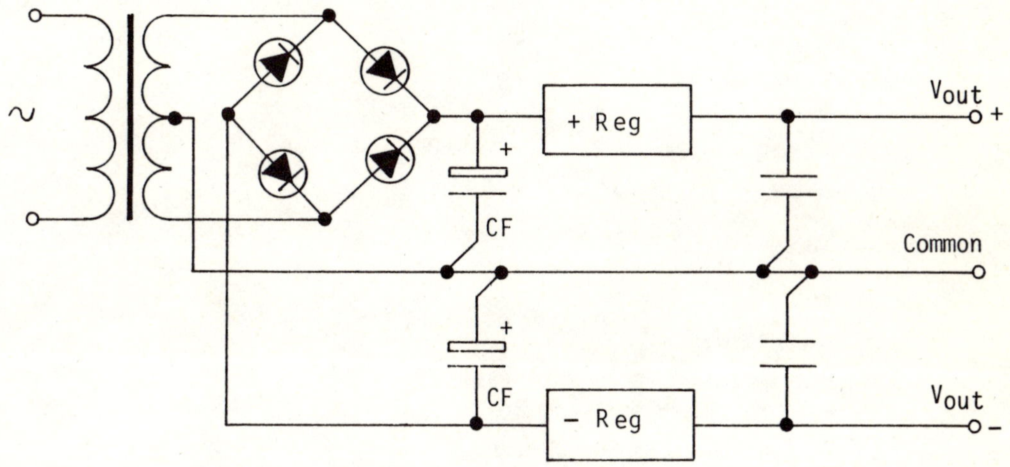

Figure 4.30 A Dual complementary supply

The choice of straight full-wave or full-wave bridge circuit depends on the voltage available at the output of the transformer. The bridge rectifier makes the best use of the transformer, but it requires 4 diodes and hence a voltage drop across the diodes of approximately 2 volts. A centre-tapped full-wave circuit loses only 1 volt across the diode and so may be preferred in low voltage supplies.

The other component to be considered, is the filter capacitor, C_F. A specification rarely considered with electrolytic capacitors used in power-supply filtering is the heating factor. With low values of D.C. current the peak-to-peak ripple voltage usually determines the size of the capacitor. However, at currents in excess of 1.5 amperes, a fair amount of heating takes place inside the capacitor itself. Manufacturers put temperature ratings on their capacitors; for example, a rating of

1000 μF/63 V at 65°C is capable of handling approximately 3 amperes (R.M.S.). As the ambient temperature increases the current capability of the capacitor derates.

Capacitors are the number one cause of power supply failure, often due to a hostile operating environment. Careful consideration when choosing a value of filter capacitor can prevent your power supply from becoming a statistic.

4.10 SELF-EVALUATION QUESTIONS (ANSWERS IN APPENDIX A)

1. Why is an input capacitor required on the regulator when the regulator is more than 5 cm away from power supply filter capacitor?

2. (a) Is an output capacitor essential for stable regulator operation?

   (b) What is the purpose of it when included in the circuit?

3. What are the basic things to consider when heatsinking a regulator I.C.?

4. Name three factors that finally determine how efficient and reliable the regulator operation is going to be. Briefly explain what is meant by each.

5. As the operating temperature of the regulator increases, what happens to the output current capability? Why does this happen?

6. How can a regulator be protected against a shorted input? Can this protection serve any other purpose?

7. Draw a basic positive regulator with input and output decoupling and full protection.

8. How can the output of a fixed three-terminal positive regulator be made adjustable?

9. Is it true to say that the external circuitry for a negative regulator is the same as that for a positive regulator? Explain your answer.

10. Draw a basic dual-polarity power supply including the transformer, rectifier, and filter components.

11. What is meant by a 'tracking regulator', and what are its advantages over a standard dual circuit?

12. What is a typical value for the 'drop-out' voltage of an integrated circuit regulator?

13. (a) A 5 volt regulator is to draw a steady load current of 1.0 A. If the input voltage is 45 volts, what is the power dissipated by the regulator, and how will this affect it's operation?

    (b) How can this disadvantage be remedied?

5 Voltage Multipliers

5.1 INTRODUCTION

Voltage multiplication circuits are used to develop D.C. voltages with a higher value than the peak A.C. input voltage without the use of a step-up transformer. These circuits can double, triple, or quadruple the input voltage if desired, but they are only able to supply a small amount of current. A television picture tube is an example of a device requiring a high value of D.C. voltage. The current requirement is small, and therefore the high D.C. voltage can be supplied more economically by means of a voltage multiplier circuit than by a transformer.

The basic half-wave rectifier circuit (as discussed in Chapter 1) can be used as a simple voltage multiplier circuit, with careful selection of a CR circuit. Figure 5.1 shows a typical circuit with a capacitor across the output, i.e. a capacitor input filter.

The transformer in Figure 5.1 has a 1 : 1 turns ratio and is used to isolate the circuit from the A.C. mains. In some countries the mains input is fed directly to Points A and B. This is called a 'transformerless supply', and it is potentially dangerous for the technician. The secondary voltage, therefore, is in this case 240 volts R.M.S. When Point A is positive with respect to Point B, the diode will conduct, and current passes through the resistor. The capacitor will charge up to the peak value of 339.4 V (i.e. 1.414 × 240).

On the next half cycle of input, Point A will be negative with respect to Point B, and the diode will not conduct. The capacitor will discharge through the resistor R. This action repeats itself every half cycle when the diode is forward biased, and it results in a varying D.C. voltage across R. Each time the diode conducts, C charges to the peak value of V_s; however, the extent to which C discharges on the negative half cycle of input is determined by the output time constant (C × R).

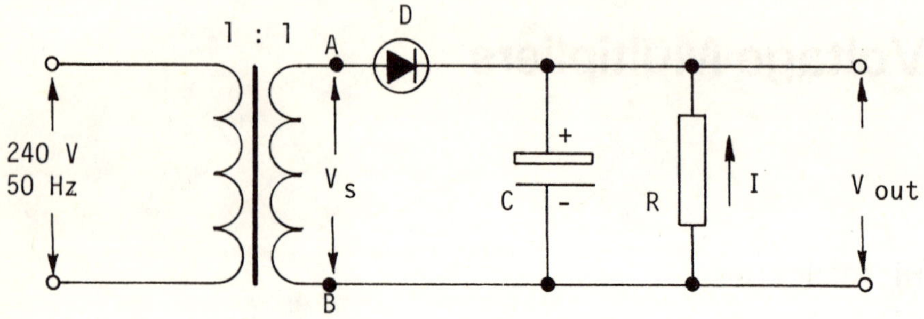

Figure 5.1 Simple half-wave rectifier with CR network across the output

Figure 5.2 shows the output waveforms for three different time constants. With 240 volts applied across V_s, it is possible to obtain a D.C. output

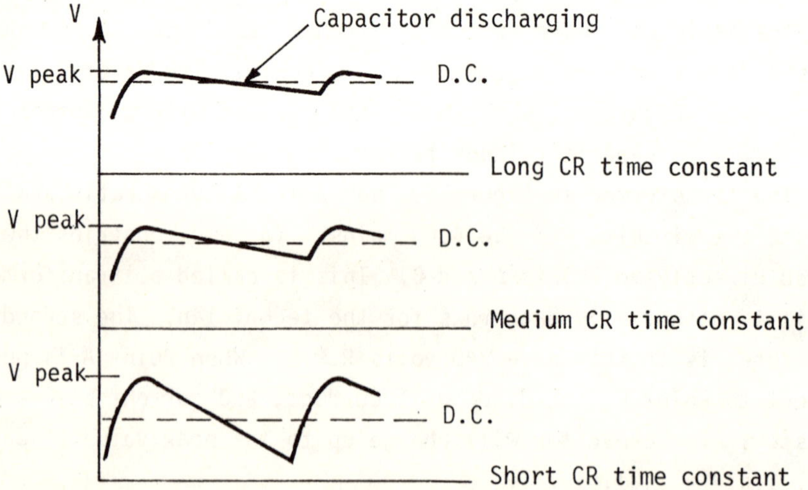

Figure 5.2 Output waveforms with three different time constants

voltage greater than the input by using a rectifier and a suitable CR circuit.

5.2 VOLTAGE DOUBLERS

5.2.1 HALF-WAVE VOLTAGE DOUBLER

A modification to the capacitor input filter allows us to build up an output voltage that may be two, three, or four times the secondary voltage. Figure 5.3 shows the circuit of a half-wave voltage doubler, often referred to as a 'cascade voltage doubler'.

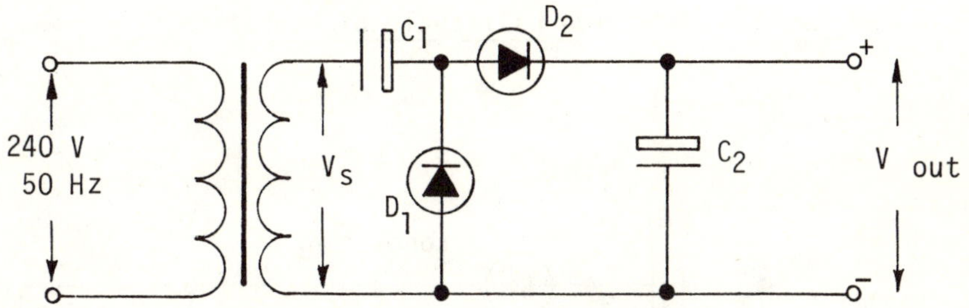

Figure 5.3 A half-wave voltage doubler

The value of the secondary voltage (V_s) is determined by the turns ratio of the transformer, and the operation of the circuit is best understood by looking at each half cycle of input separately. Figure 5.4 shows the circuit with 240 V 50 Hz input and a turns ratio of 1 : 1. During the negative half cycle of input voltage, Point A of the transformer secondary is negative with respect to Point B. Diode D_1 is forward biased, and the current path is in the direction of the arrow. C_1 charges to the peak value of V_s (i.e. 339.4 V) with the polarity shown. (There is very little forward voltage drop across a good diode, and for simplicity we will ignore it.)

On the next half cycle of input voltage, the polarity of V_s reverses. In Figure 5.5 Point A is now positive with respect to Point B. Diode D_1 is cut off by the positive voltage on its cathode. Diode D_2 has the secondary voltage (V_s) in series with the voltage stored in C_1 applied to its anode. D_2 conducts and the current path is in the direction indicated, charging C_2 to 2 V_s.

$$V_{C2} = 339.4 \text{ V} + 339.4 \text{ V} = 678.8 \text{ V}$$

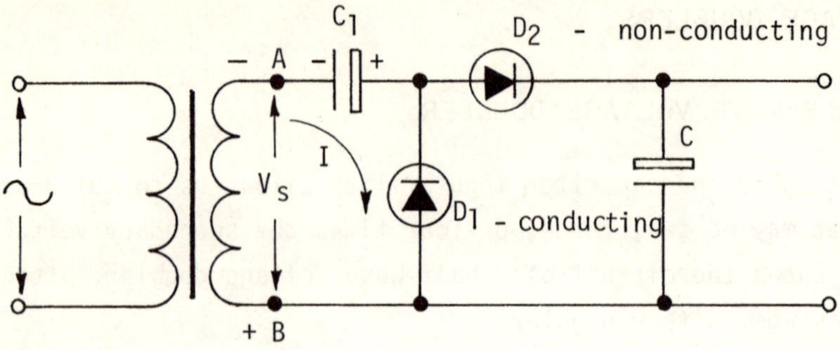

Figure 5.4 Negative half cycle of operation

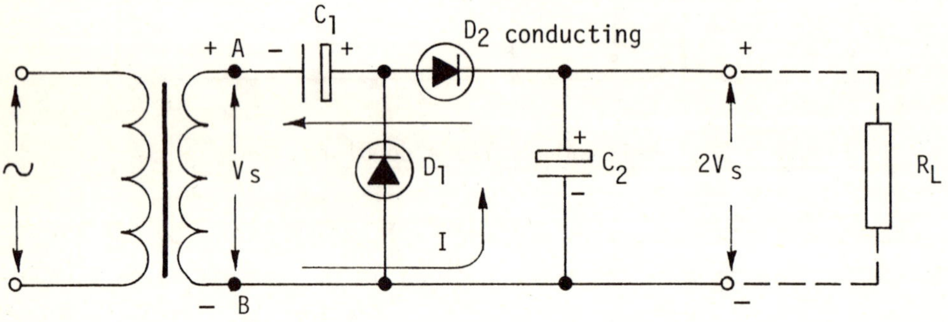

Figure 5.5 Positive half cycle of operation

If no load is connected, capacitor C_1 will stay charged to V_s and capacitor C_2 to 2 V_s. However, the idea of the circuit is to supply voltage to a load. Capacitor C_2 discharges through the load on the next negative half cycle and is recharged to 2 V_s during the positive half cycle. An output voltage of reverse polarity can be obtained by simply reversing the diodes and the capacitors.

The output waveform across capacitor C_2 is the same as that of a half-wave rectifier with a capacitor filter. The peak inverse voltage across each diode is 2 $V_{s\ peak}$, and the ripple frequency is the same as the input.

5.2.2 FULL-WAVE VOLTAGE DOUBLER

Another doubler circuit uses the full-wave principle. Two capacitors are individually charged, and the output is the sum of the two voltages. Figure 5.6 shows a full-wave voltage doubler circuit.

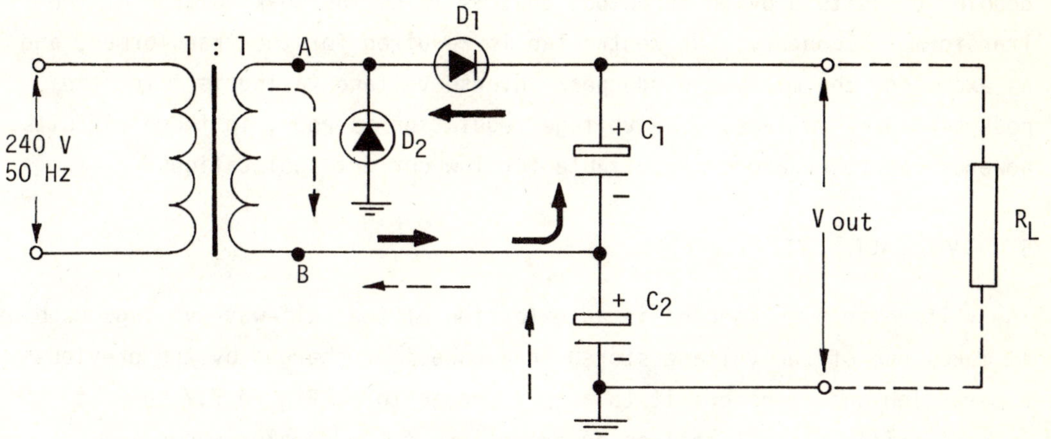

Figure 5.6 A full-wave voltage doubler

When Point A is positive with respect to Point B, diode D_1 conducts. Current is in the direction shown by the heavy arrows, and capacitor C_1 charges up to 339.4 V_{peak} (1.414 × 240); this is V_{C1}. When Point A is negative with respect to Point B, diode D_2 conducts. Current follows the path of the dotted arrows, and capacitor C_2 charges up to 339.4 V_{peak} (1.414 × 240); this is V_{C2}.

The output voltage V_{out} is therefore the sum of the voltages stored in C_1 and C_2.

$$V_{out} = V_{C1} + V_{C2}$$

$$V_s = 1.414 \times 240 \text{ V}$$

$$= 339.4 \text{ V}_{peak}$$

Therefore $$V_{out} = 339.4 \text{ V} + 339.4 \text{ V}$$

$$= 678.8 \text{ V}_{peak}$$

When current is drawn by the load, the voltage across capacitors C_1 and C_2 is the same as that across a capacitor filter in a full-wave rectifier circuit. However, because C_1 and C_2 are in series, the total capacitance is less, and this results in a poorer filtering action.

The peak inverse voltage across each diode is $2 \times V_{s\ peak}$, and the ripple frequency is twice the input. Both the half- and full-wave voltage doubler circuits provide an output that is twice the peak voltage of the transformer secondary. No centre-tap is required for the transformer, and as expected, the minimum diode peak inverse voltage rating is twice the peak secondary voltage. The voltage regulation is poor, in these circuits, however, and they are only suitable for low current applications.

5.3 VOLTAGE TRIPLER

The voltage tripler concept is an extension of the half-wave voltage doubler. It makes use of the voltage stored in a capacitor charged by the previous alternation and transfers it to a third capacitor. Figure 5.7 shows a voltage triple-circuit that is an extension of the doubler shown in Figure 5.3.

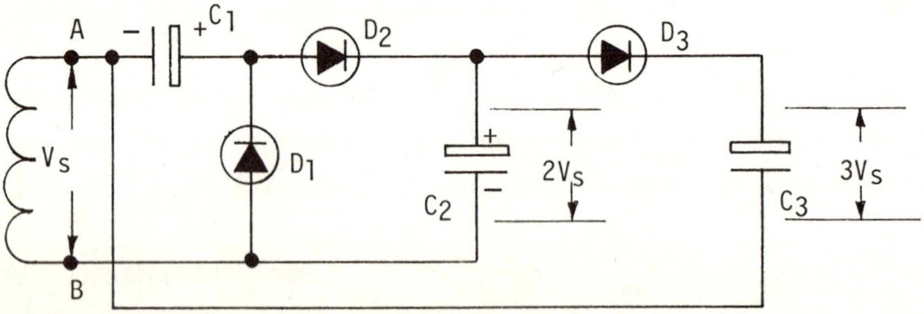

Figure 5.7 Voltage tripler

On the negative half cycle of input, capacitor C_1 charges to V_s, with the polarity shown. When D_2 is forward biased on the next half cycle (positive), the voltage stored in C_1 is series-aiding the input voltage V_s. A voltage approximately equal to 2 V_s is then transferred and stored in C_2 with the polarity as shown in Figure 5.7. So far, this is the same as the action of the voltage doubler circuit in Figure 5.3.

On the next negative half cycle of input, Point B is positive with respect to Point A. Therefore, V_s is now series-aiding the voltage stored in C_2, as shown in Figure 5.8. The positive potential on D_2 cathode biases the diode off. The anode of D_3 is at a potential of $2V_s$ and is therefore forward biased. Current follows the direction of the arrow, and C_3 charges to $3V_s$.

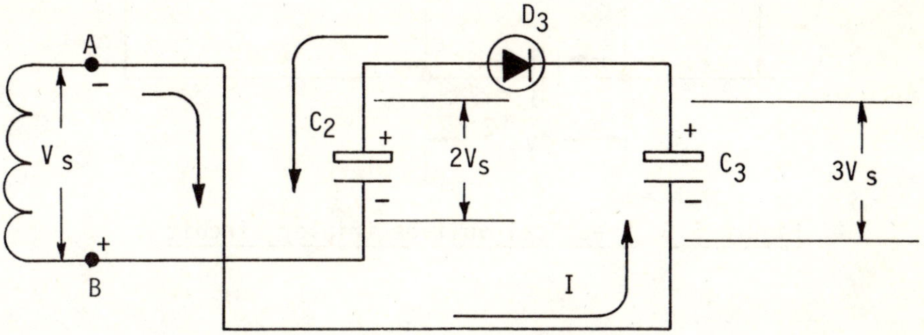

Figure 5.8 V_s in series with V_{C2}

At the same time, diode D_1 (as shown in Figure 5.7) is forward biased due of the negative polarity at Point A, and capacitor C_1 recharges to V_s. This prepares it for the next positive half cycle.

The practical approach is to charge the third capacitor to 2 V_s and tap the triple voltage off across two capacitors in series ($V_s + 2V_s = 3V_s$). Figure 5.9 shows this type of circuit arrangement.

On the negative half cycle of input voltage, capacitor C_1 charges to V_s peak through diode D_1. Capacitor C_2 charges to twice the peak voltage (the sum of $V_s + V_{C1}$) during the positive half cycle of input voltage. During the next negative half cycle D_3 conducts, and the voltage across C_2 charges C_3 to the same value of $2V_s$, as shown in Figure 5.10. At the same time D_1 is again forward biased, and capacitor C_1 is recharged.

In summary then, after switch on D_1 conducts on the first negative half cycle charging C_1 to V_s. On the first positive half cycle D_2 conducts. The charge in C_1 is in series with V_s, and C_2 charges to $2V_s$. During the second negative half cycle, D_1 conducts 'topping up' capacitor C_1; diode D_3 conducts and C_2 charges C_3 to $2V_s$. The output voltage is taken across capacitors C_1 and C_3 in series, and it equals three times the input voltage.

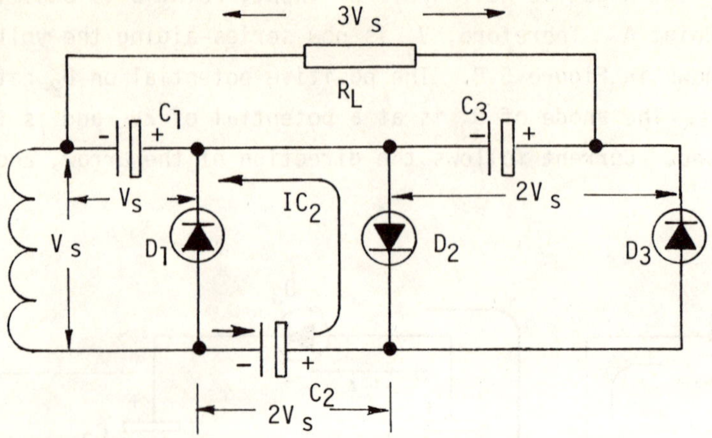

Figure 5.9 Practical voltage tripler circuit

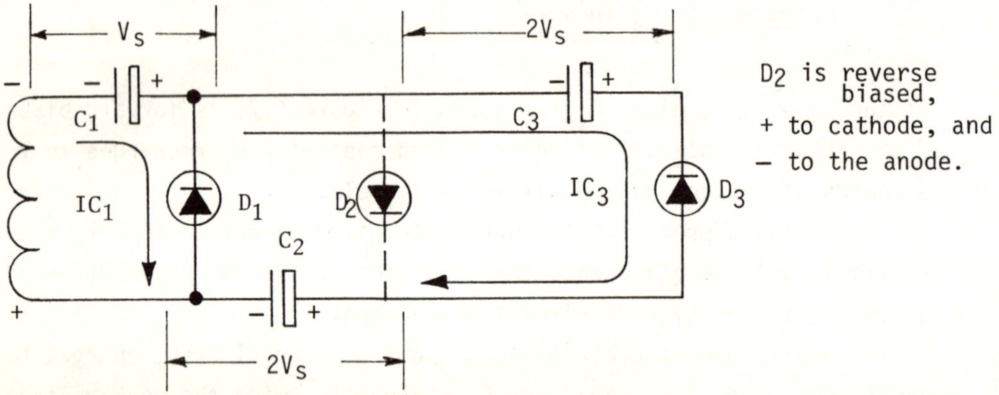

Figure 5.10 Current paths on the second negative half cycle

If V_s = 240 V A.C., then the peak value is 339.4 V.

$$V_{out} = V_{C1} + V_{C3}$$

$$= 339.4 \text{ V} + 678.8 \text{ V}$$

$$= 1018.2 \text{ V} = 3V_s$$

Let us now follow a bit further the operation of the voltage tripler circuit just described. Assume the first *complete cycle* of operation has passed. Capacitor C_1 charges to V_s, and this series-aids V_s to charge capacitor C_2 to 2 V_s.

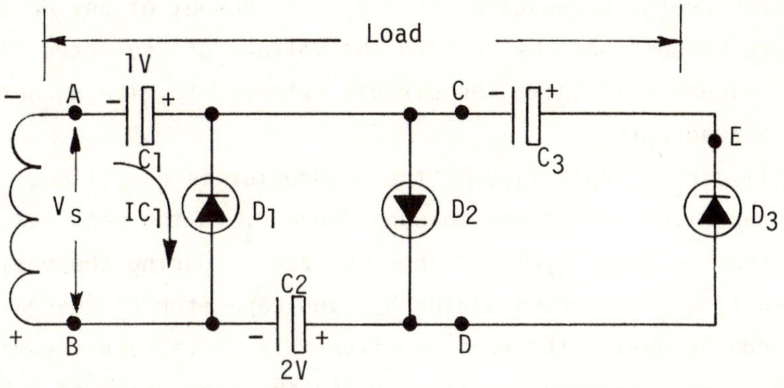

C_1 charged to V_s

C_2 charged to $2V_s$

Figure 5.11 Second negative half cycle of input

At the commencement of the next negative half cycle, Point A is negative with respect to Point B (see Figure 5.11). Two things now happen simultaneously: Diode D_1 is forward biased and C_1 'recharges'. The charge from capacitor C_2 is transferred to capacitor C_3 in the following way. (For the purpose of explanation, let us assume that V_s = 1 volt; then V_{C1} = 1 volt and V_{C2} = 2 volts). Point C (anode of D_2) is 1 volt positive with respect to Point A because of the charge in C_1. The input voltage V_s is series-aiding the voltage across capacitor C_2. Therefore, Point D is 3 volts positive with respect to Point A. The 3 volts on D_3 anode forward biases the diode, and therefore Point E is also at a potential of 3 volts.

Capacitor C_3 now has 1 volt at Point C and 3 volts at Point E. It will charge to the difference (3-1 = 2 volts) or the equivalent to 2 V_s. As current is drawn by the load, the voltage across C_1 drops slightly, forward biasing D_1 and allowing it to conduct. This recharges C_1.

It takes one cycle of operation (approximately 20 milliseconds) to set the circuit up, i.e. C_1 charges to V_s and C_2 charges to 2 V_s. After that, diodes D_1 and D_3 conduct on the negative half cycle of input, charging C_1

and C_3. Diode D_2 conducts on the positive half cycle of input, charging C_2.

5.4 VOLTAGE QUADRUPLER

It is now apparent that by means of the addition of a diode/capacitor network we can charge a capacitor to 2 V_s. An output of any multiple of V_s can therefore be obtained, by tapping the voltage off selected capacitors in series. Figure 5.12 shows the circuit extended to give an output of 4 V_s, i.e. a quadrupler.

The full circuit operation for the quadrupler is as follows. In operation, capacitor C_1 charges through diode D_1 to the peak value of V_s during the negative half cycle of input voltage. During the positive half cycle of input, V_{C1} is series-aiding V_s, and capacitor C_2 charges to 2 V_s. Diode D_3 conducts during the next negative half cycle, and the voltage across capacitor C_2 charges capacitor C_3 to the peak value of 2 V_s. Capacitor C_1 is also 'topped up'. On the next positive half cycle, diodes D_2 and D_4 conduct with capacitor C_3, and they charge capacitor C_4 to 2 V_s peak. This area of the circuit operation is shown in Figure 5.13. Its behaviour is exactly the same as the transfer of voltage from C_2 to C_3 in the tripler circuit. However, the input voltage is now series-aiding the voltage across C_1 and C_3. Point B is at a potential of $V_s + V_{C1} + V_{C3} = 4\ V_s$ This forward biases D_4, and therefore Point C is also at 4 V_s. Point D is at a potential of 2 V_s because of the charge in C_2, and C_4 charges to the difference of $4\ V_s - 2\ V_s = 2\ V_s$.

On the negative half cycles of operation, capacitors C_1 and C_3 charge, while on the positive half cycles of operation, capacitors C_2 and C_4 charge. Even multiples of the input voltage are taken from the bottom end of the transformer and odd multiples from the top end, as shown in Figure 5.12. One further stage is shown within the dashed lines. The inclusion of D_5 and C_5 enables an output of 5 V_s to be obtained.

The transformer rating is only V_s peak, and each diode in the circuit must have a peak inverse voltage equal to two times the peak value of the source voltage, i.e. $2 \times V_s\sqrt{2}$. With small load currents and good quality capacitors that have very little leakage, very high D.C. voltages can be obtained from this circuit configuration. Every additional stage is capable of stepping the output voltage up by 2 V_s.

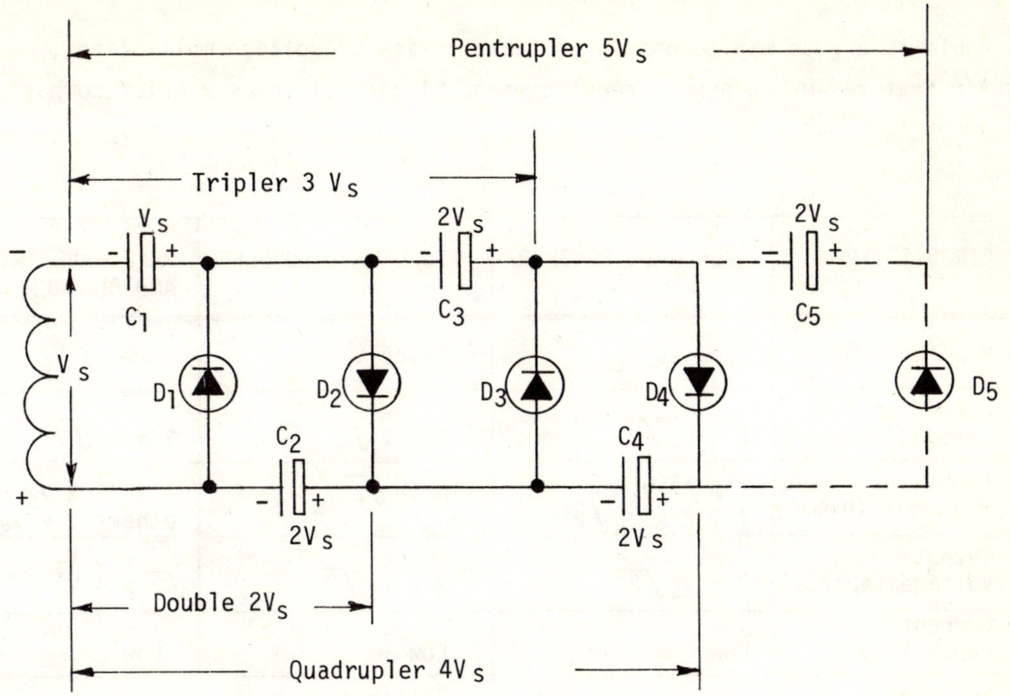

Figure 5.12 A voltage quadrupler circuit

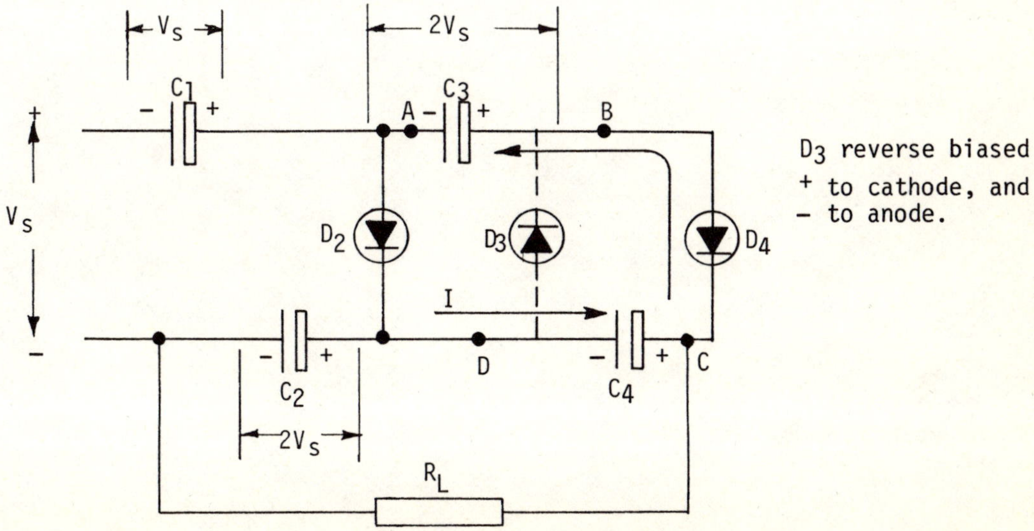

Figure 5.13 C_3 charging capacitor C_4 to 2 V_s

5.5 FEATURES OF VOLTAGE MULTIPLIERS

A table is a good way to obtain an overall view of voltage multipliers, their features and component requirements. Table 5.1 shows a brief comparison between four basic circuits.

CIRCUIT TYPE	HALF WAVE DOUBLER	FULL WAVE DOUBLER	HALF WAVE TRIPLE[R] AND QUADRUPLER
Ripple	Same as V_s	Twice V_s	Same as V_s
P.I.V. of diodes (minimum)	$2 \times V_s\sqrt{2}$	$2 \times V_s\sqrt{2}$	$2 \times V_s\sqrt{2}$
Capacitor working voltage(minimum)	1 @ $\sqrt{2}\ V_s$ 1 @ $2 \times V_s\sqrt{2}$	Both $\sqrt{2}\ V_s$	1 @ $\sqrt{2} \times V_s$ others $2 \times V_s\sqrt{2}$
Output voltage (approx)	$2 \times V_s\sqrt{2}$	$2 \times V_s\sqrt{2}$	$\sqrt{2} \times 3\ V_s$ $\sqrt{2} \times 4\ V_s$
Current capability	Low	Low	Low
Voltage regulation	Poor	Better than half wave doubler	Poor

Table 5.1 Comparison of voltage multiplier features

5.6 SELF-EVALUATION QUESTIONS (ANSWERS IN APPENDIX A)

1. Why are voltage doublers and triplers required?

2. With the aid of diagrams, explain the operation of a full-wave voltage doubler circuit.

3. What would be the specifications of the following for a half-wave voltage quadrupler circuit?

   (a) P.I.V. (minimum)

   (b) D.C.W.V. of the capacitors

   (c) Ripple frequency

4. Draw the circuit of a multiplier that will step up the input voltage by six times. Explain how each capacitor obtains its charge.

5. What is the main disadvantage of a multiplier circuit?

6. What is a practical application for a voltage multiplier?

6 D.C. to D.C. Converters

6.1 INTRODUCTION

The power supplies discussed in earlier chapters have all been designed to convert A.C. input voltage to a D.C. voltage. This voltage has then been filtered to remove any ripple and then fed to some type of voltage regulator.

There are many cases where a low D.C. voltage (12 volts from a car battery, for example) must be changed or converted to several hundred volts D.C. Basically, a D.C. to D.C. converter takes a low D.C. voltage and converts it to a high A.C. voltage. The A.C. is then rectified to provide a higher D.C. voltage.

Practical applications of this type of circuit are:

1. To power portable valve equipment.

2. To provide a high D.C. voltage for power transistor applications, e.g. transmitters. This is perhaps the most familiar application.

3. Voltage 'step-up' circuit for automobile ignition systems.

The forerunner in D.C. to D.C. converters was the vibrator power supply, which was followed soon after by a transistorized unit.

6.2 VIBRATOR POWER SUPPLIES

Vibrator power supplies evolved from the simple circuit shown in Figure 6.1. The magnetic field set up in the exciter coil pulled the points apart against a fixed spring tension. This sent pulses of current through the primary winding of a transformer, which was then stepped up to a higher voltage by the transformer turns ratio. The output waveform is effectively half-wave.

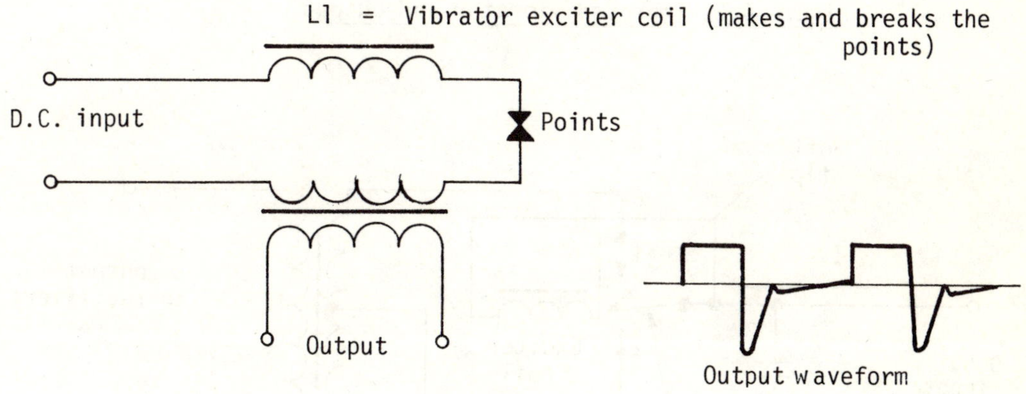

Figure 6.1 Basic vibrator circuit with output waveform

The first improvement to the basic circuit was to obtain a full-wave balanced output. This is achieved by centre-tapping the output transformer and switching the input across it, as shown in Figure 6.2. The result approximates a square wave.

The following explanation of a vibrator power supply is based on Figure 6.3. When the D.C. power is applied, current passes through the exciter coil (which is an electromagnet) and down through the top of the output transformer primary to the positive side of the supply. The magnetic field pulls the soft iron tip towards the core, and closes Contact 1. This sends a 'pulse of current' through the top of the primary winding. When Contact 1 is closed, there is a short circuit across the exciter coil. With no current through the exciter coil, there is no magnetic field, and the vibrator arm falls back, striking Contact 2. This sends a 'pulse of current' through the bottom of the primary winding. With the short circuit

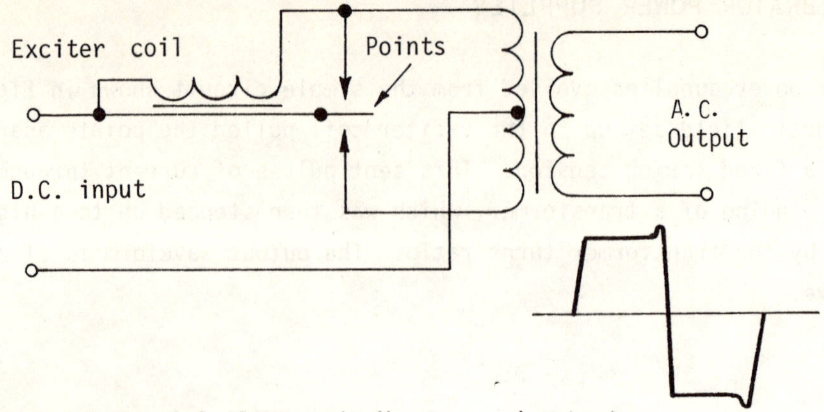

Figure 6.2 Improved vibrator and output

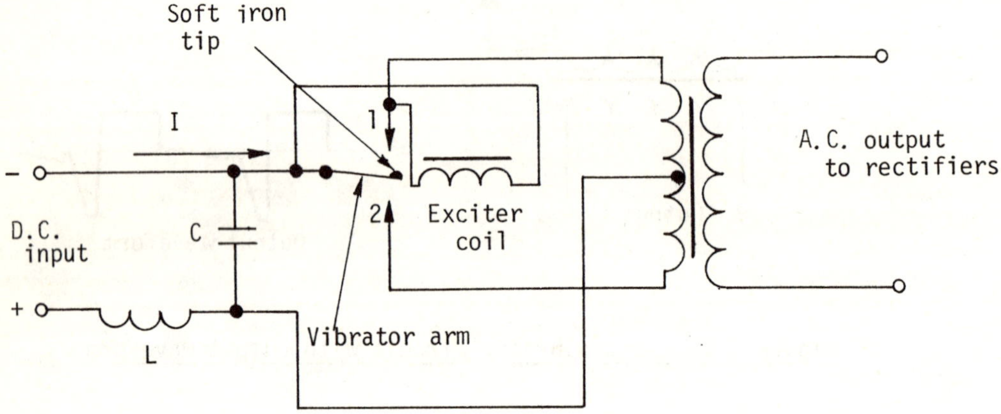

Figure 6.3 Vibrator circuit

across the exciter coil now removed, the electromagnetic field builds up and again attracts the soft iron tip. Contact 2 opens and Contact 1 closes, resulting in another 'pulse of current' through the top of the primary.

The action repeats itself. Two opposite-going pulses are seen by the secondary, and these are stepped up by the turns ratio of the transformer. The A.C. variation, which is now present in the secondary circuit, is fed to a rectifier circuit which converts it to a D.C. voltage at a higher level than the original D.C. input.

Any switching circuit in which contacts are making and breaking generates noise. Inductor L is a 'radio frequency choke' (R.F.C.) and, together with the capacitor, it reduces transient spikes. They also prevent radiation of high frequency signals and help prevent the surface of the points from pitting.

Vibrators have limited life; also, the units are bulky, generate spark interference and are not very efficient at low voltages. Typical output voltages for a vibrator power supply to be used with valve equipment are 150 to 250 volts. With the advent of transistorized equipment which operated at a much lower voltage, it was not surprising to see the vibrator replaced by a transistor converter.

6.3 TRANSISTOR CONVERTERS

In the vibrator converter, the switching action was carried out mechanically. In a transistor converter, the transistor or transistors behave as electronically controlled 'on-off' switches. The transistor can provide a useful switch. When it is off, only the collector cut-off current flows. When it is 'on', it is saturated, and behaves as a low value of resistance. This action forms the basis of all transistor converters, although there are several possible circuit configurations. The most common types are:

1. Transformer coupled, single transistor

2. Push-pull transformer coupled, two transistors

The push-pull circuit is the most efficient and popular.

The type of transistor used for converter application should be capable of handling high current, and should have a low saturated resistance, a high maximum collector voltage rating, and an adequate dissipation rating. Transistor D.C. converters can usually handle about three to ten times as much power as the transistor dissipation rating.

6.3.1 TRANSFORMER COUPLED, SINGLE TRANSISTOR

This circuit is probably the simplest D.C. converter available; it consists of a single transistor connected to a basic transformer. The circuit is shown in Figure 6.4.

The base of transistor Q_1 is slightly forward biased by resistors R_1 and R_2. This allows a small emitter - collector current to flow. This small current through L_1 induces a potential in L_2, which increases the forward bias on transistor Q_1, causing a further increase in collector current. This action continues until saturation point is reached and Q_1 collector voltage is minimum. At this point no further change in current takes place; the magnetic field collapses, reversing the polarity induced

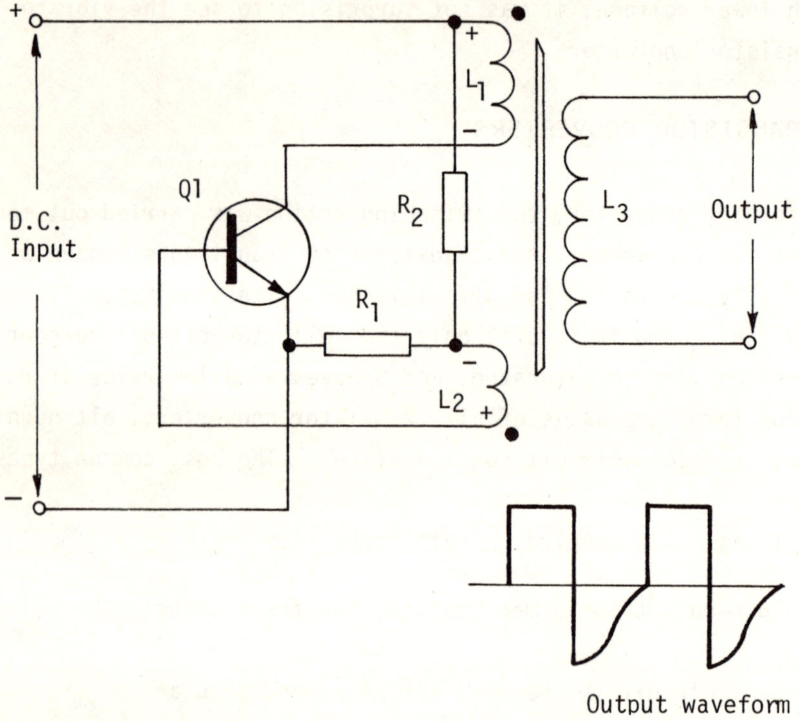

Figure 6.4 Transformer coupled single transistor D.C. converter

in L_2. This drives the base of the transistor negative, which turns the transistor off and cuts off the collector current. Once the energy from the collapsing field is spent, the transistor is forward biased again by R_1 and R_2, and the process then repeats itself. The 'switched' pulses of D.C. are stepped up by the secondary winding of the transformer (L_3) and rectified.

The operating frequency is determined by the number of turns on the primary and by the size and permeability of the core. The most efficient frequency is approximately 1000 hertz.

As with the basic vibrator circuit, the output is effectively half-wave, and therefore the efficiency of the circuit is not very high. Output power is usually less than 10 watts.

6.3.2 PUSH-PULL TRANSFORMER COUPLED

To illustrate the operation of a push-pull transistor D.C. converter, a basic circuit with no biasing components is shown in Figure 6.5. In addition to the centre-tapped primary and the secondary winding, the transformer has two feedback windings. The primary and the two feedback windings are wound together, turn for turn (that is, 'bifilar wound'). This ensures that both windings are identical with respect to inductance and resistance, and also gives the added benefit of unity coupling. With unity coupling, there is minimum overshoot in the induced voltage when the magnetic field collapses.

If a standard transformer winding technique is used, the transistors must be able to withstand a higher collector voltage to enable them to handle the switching transients.

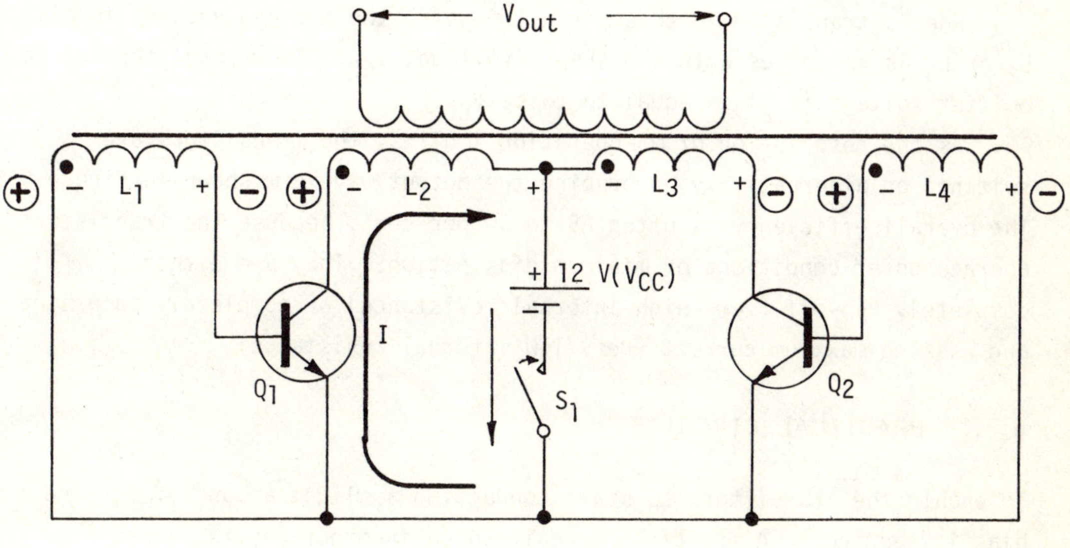

(the dots indicate the start of the windings)

Figure 6.5 Basic push-pull D.C. converter circuit

Transistors Q_1 and Q_2 in Figure 6.5 should be a similar type N.P.N., which require a positive potential on the base to turn them on. When switch S_1 is closed, current will initially flow through both Q_1 and Q_2. However, the two transistors will have slightly different characteristics, making one transistor conduct harder than the other and take over the circuit operation. If Q_1 conducts harder, current passes through L_2 in the direction indicated. The rising current will produce a rising magnetic field, which will induce voltages in L_3 and the feedback windings, with the polarities shown within the circuit. The base of Q_2 is driven negative and therefore turns off; the base of Q_1 is driven positive, and the transistor goes into saturation. Thus, transistor Q_1 is 'on' and the transistor Q_2 is 'off'.

At this point, no further change in current takes place. The electromotive force induced in L_1 drops, and as the forward bias on Q_1 is reduced, Q_1 collector current decreases. The magnetic field around L_2 collapses, which induces a voltage in the windings with the opposite polarity (shown within the circles in Figure 6.5). Transistor Q_1 is turned off hard by the negative base voltage, and at the same time transistor Q_2 is turned on by the positive base voltage and thus goes into saturation. Now Q_2 is 'on' and Q_1 is 'off'. The process then repeats itself.

When a transistor in this circuit is off, the induced voltage in either L_2 or L_3 is in series with the supply voltage, V_{CC}. Therefore, the collector-emitter voltage (V_{CE}) is equal to twice V_{CC}.

As the description of the operation implies, the transistors are switched on alternatively, producing the output waveform shown in Figure 6.6. The overall efficiency is often 85 to 90 per cent, because the transistors operate under conditions of minimum dissipation. They are either completely cut-off (very high internal resistance) or completely saturated and passing maximum current (very low internal resistance).

6.3.3 PRACTICAL CIRCUITS

To enable the transistors to start conducting a slight amount of forward bias is required. A practical circuit which incorporates the biasing components is shown in Figure 6.7.

The primary circuit is a push-pull 'Armstrong' oscillator with two 5-turn feedback coils and a 40-turn primary. When the circuit oscillates on the first half cycle, Q_1 draws current down through the top half of

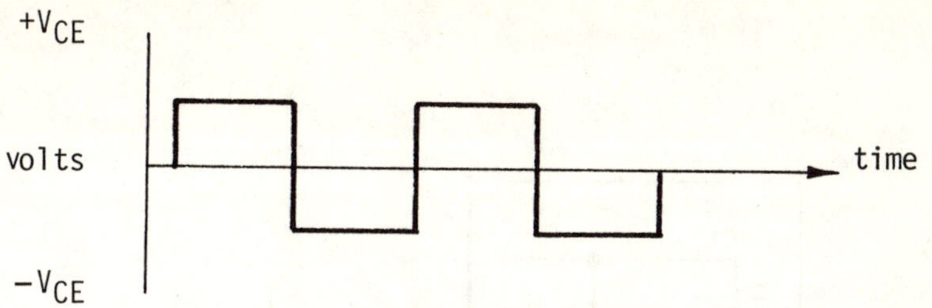

(a) Collector-emitter waveform for transistors

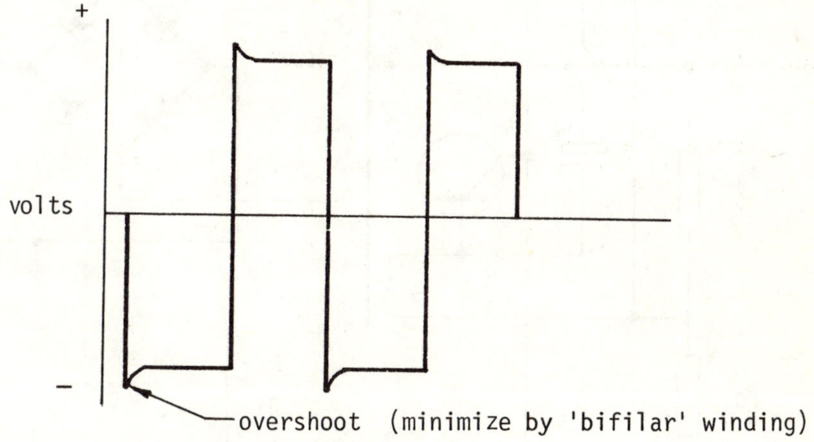

(b) Output waveform from transformer secondary, (stepped-up)

Figure 6.6 Output voltage waveforms for a push-pull converter

the primary. On the next half cycle, Q_2 draws current up through the other half of the primary.

The secondary 'sees' the primary current as alternating and steps up this A.C. voltage. The high A.C. voltage is then rectified and filtered. Each transistor is driven into saturation on alternate half cycles. This results in a flat-topped secondary waveform which is easier to filter. Care must be taken to ensure that the transistors and diodes can handle possible high-voltage transients. Winding the primary bifilar will reduce transients. Capacitors C_1 and C_2 present a reactance of approximately 12.7 Ω at the operating frequency of 500 Hz and help protect Q_1 and Q_2 input circuit.

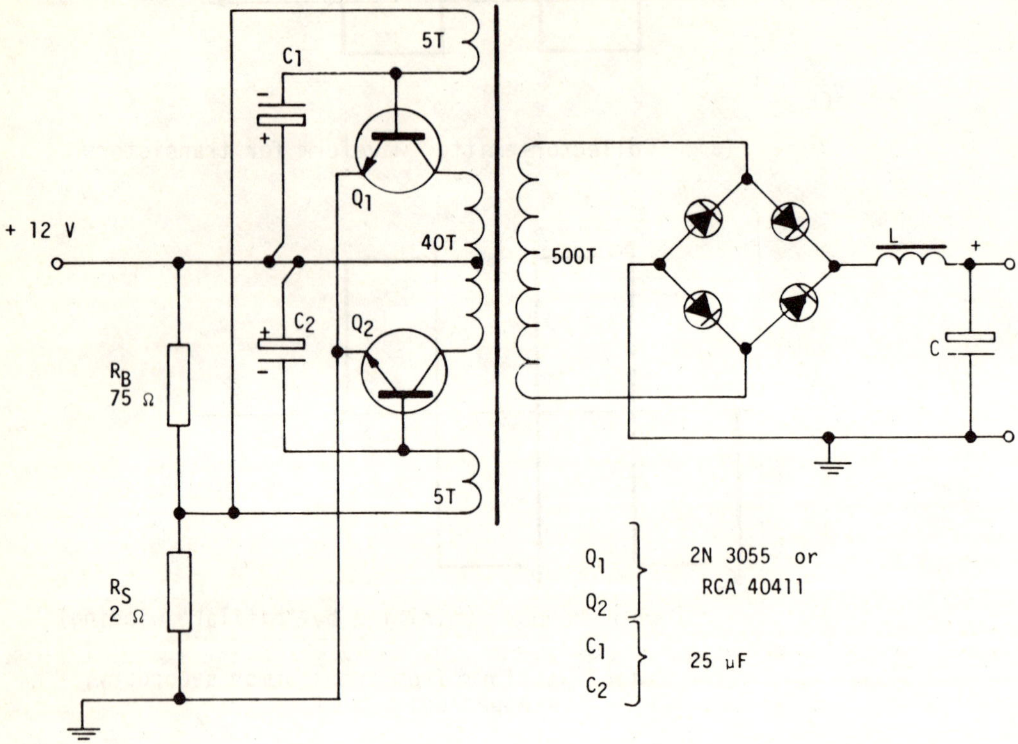

Figure 6.7 Practical converter circuit (with full-wave rectification and LC filter

This circuit should produce about 40 watts of D.C. power. If a higher power is required, the oscillator transistors should be upgraded and R_B and R_S reduced in proportion to increase current flow.

Unlike other power supply circuits, the harder a push-pull D.C. converter works, the more efficient it becomes. Suppose that for the circuit in Figure 6.8 Input power = 12 V @ 4.3 A = 51.6 watts. If the output voltage is 300 V @ 100 mA, then output power = 300 × 100 mA = 30 watts.

$$\text{Efficiency} = \frac{\text{Power out}}{\text{Power in}} \times 100\ \%$$

$$= \frac{30}{51.6} \times 100$$

$$= 58.1\ \%$$

If the output current is increased by decreasing the value of the load resistance to, say, 133 mA, and the input current increases to 4.6A, then the new input power will equal 12 × 4.6 = 55.2 watts.

$$\text{The output power} = 300 \times 133\ \text{mA.}$$

$$= 40\ \text{watts}$$

$$\text{Efficiency} = \frac{40}{55.2} \times 100$$

$$= 72.4\ \%$$

Therefore, as the load increases (within the limitations of the circuit design), the overall efficiency increases.

6.4 REVERSE POLARITY PROTECTION

If a D.C. converter is inadvertently connected to the supply voltage with the wrong polarity, disaster results. Suppose that in Figure 6.7 the input is connected to -12 volts; as a result both transistors will have the output forward biased. This will cause excessive current and could easily damage the transistors. Connecting a diode in its forward bias mode in series with the D.C. input will prevent this from happening. If the input voltage is of the wrong polarity, the diode is reverse biased and no current will flow. Figure 6.8 shows two other D.C. converter circuits which both have reverse polarity protection.

6.5 APPLICATIONS

Transistor D.C. converters have many uses and are particularly suitable for mobile and portable radio communications equipment. They can also supply a high A.C. voltage to operate small A.C. electric motors and fluorescent

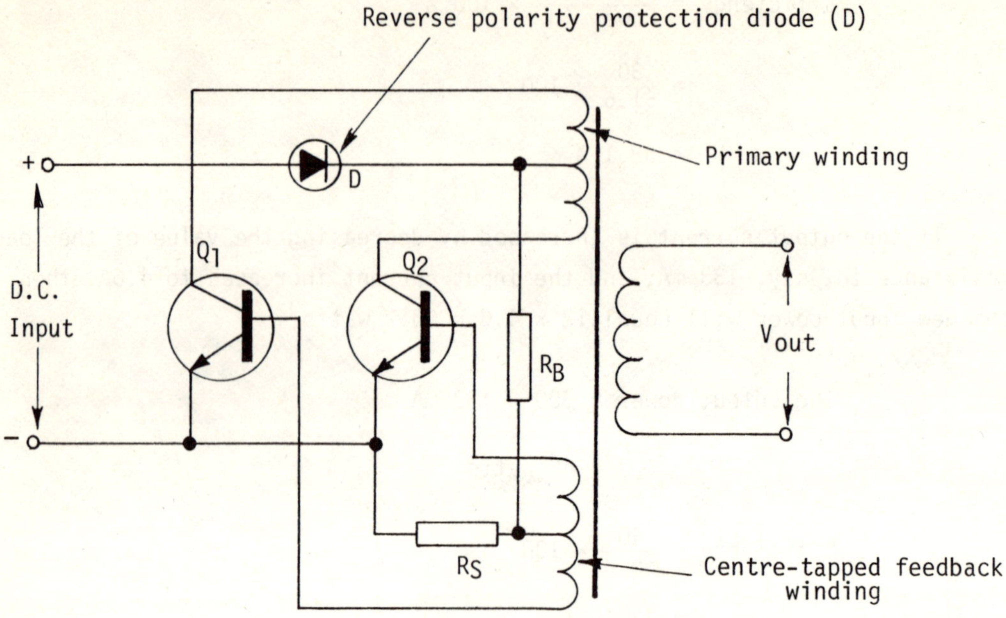

a)

Figure 6.8a Other converter circuits with reverse-polarity protection

lighting from D.C. power supplies. A transistor converter is practically instantaneous in starting, and therefore makes a valuable emergency power unit in the event of a mains supply failure.

Electronic flashes for photographic use are powered by D.C. to D.C. converters. Figure 6.9 is a typical electronic flash circuit.

The converter is powered from a low voltage source, 6 to 12 volts D.C. An output voltage of approximately 300 volts is used to charge a large electrolytic capacitor, C_2 2000 μF. When the trigger contacts in the camera are closed by the shutter mechanism, the flash is triggered. The capacitor discharges through the flash tube to give a high intensity light output.

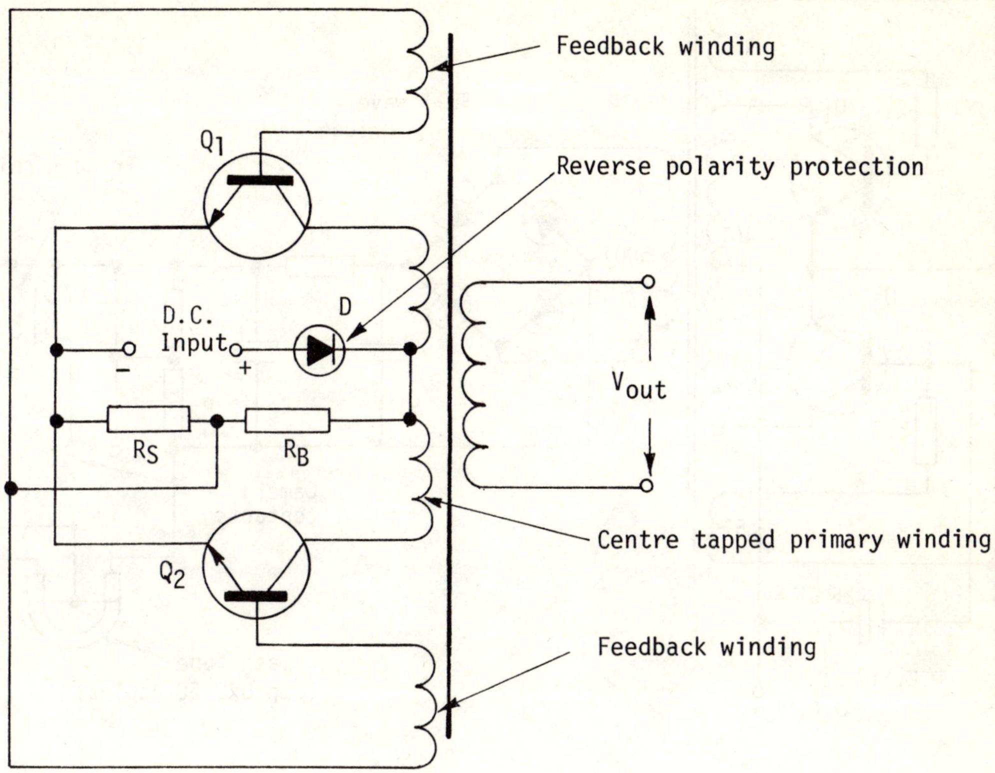

b)

Figure 6.8b Other converter circuits with reverse polarity protection

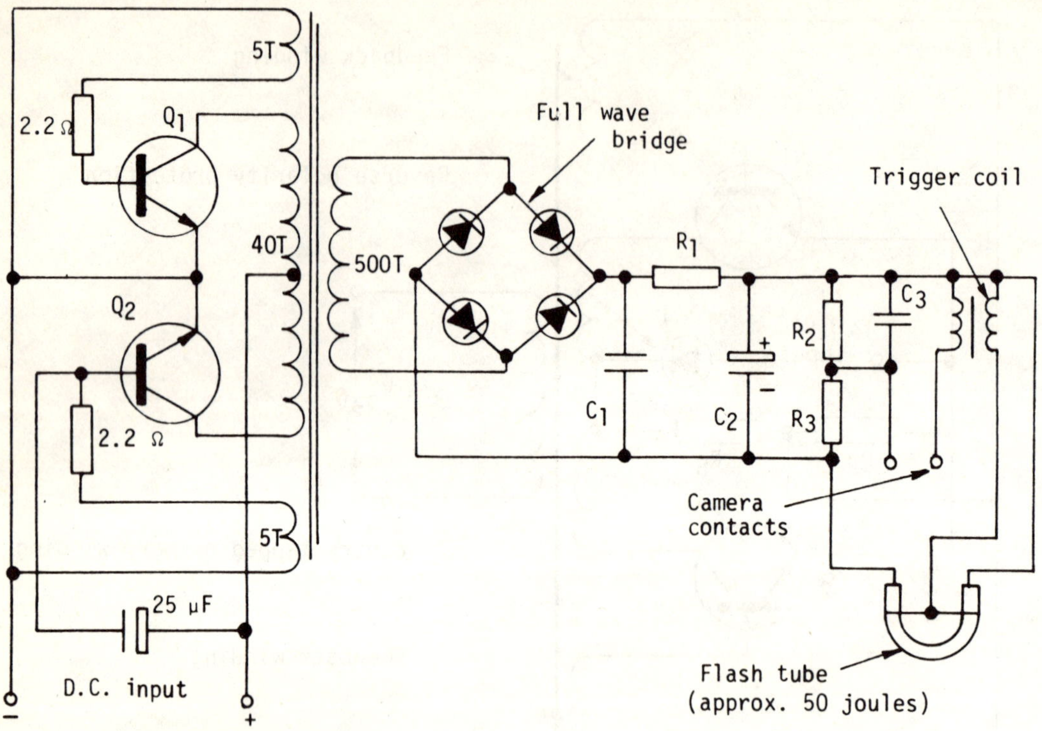

Q_1 and Q_2 = 40411 RCA

R_1	=	3.3 kΩ	C_1	=	100 nF
R_2	=	100 kΩ	C_2	=	2000 μF 350 V.W.
R_3	=	1 MΩ	C_3	=	270 nF

Figure 6.9 Transistor electronic flash

6.6 SELF-EVALUATION QUESTIONS (ANSWERS IN APPENDIX A)

1. What is the main purpose of a D.C. to D.C. converter?

2. What is the main disadvantage of a vibrator power supply at low voltage?

3. What purpose does the transistor serve in a converter circuit?

4. Briefly explain the operation of a push-pull converter circuit.

5. How can 'overshoot' in the induced voltage be minimized?

6. If the input power to a converter was 50 watts, and the output power was 44 watts, what is the efficiency? Would this be a realistic figure for the circuit in Figure 6.7?

7. How can the converter be protected from reverse input polarity?

8. What is a major operating difference between an ordinary power supply output and a converter output?

7 Switched-Mode Power Supplies

7.1 INTRODUCTION

The series-pass transistor in a conventional series regulator operates as a variable resistance which keeps the output voltage constant despite changes in input voltage and load current. This transistor must be able to dissipate the voltage difference between the input and output voltages at the load current. If the input-output voltage difference is large, or if the input voltage has a high ripple content, the power generated can be excessive. This power is radiated as heat, and extensive heatsinking is therefore required.

During the past couple of years or so, another type of regulator circuit has arrived on the scene which overcomes these disadvantages. This new type of circuit is called either a 'switched-mode power supply' (S.M.P.S.) or a 'chopper' power supply and can be encountered in either discrete or integrated format. The switching-mode type of regulator is basically a D.C. to D.C. converter that operates at high efficiency.

The transistor switches from saturation to cutoff, and therefore very little voltage is dropped across it. Saturation voltage (V_{sat}) across the transistor is very low, and the power dissipated by the transistor is greatly reduced. This is the main advantage over linear voltage regulators.

Heatsinking is kept to a minimum because the device runs much cooler than high current regulators using series-pass transistors. High efficiency operation is maintained even with quite large input-output voltage differences. In fact the voltage difference has very little effect on the efficiency.

Additional advantages of this type of regulator include the fact that the output voltage can be

(i) Stepped-up

(ii) Stepped down

(iii) or Inverted

One of the disadvantages of switching regulators is that they are more complex than linear regulators. However, this increase in electronic complexity must be weighed against the thermal and design complexity of high-current linear regulators. A further disadvantage is higher output ripple amplitude, but this can be reduced with adequate filtering. Lastly, switching regulators throw current transients back into the unregulated supply which are somewhat larger than the maximum load current. These can be troublesome unless reduced by filtering.

(Note: basic circuit concepts used in this chapter are reproduced by kind permission of National Semiconductor from their LM3524 Pulse Width Modulator publication.)

7.2 SWITCHING VERSUS LINEAR REGULATORS

Most power supplies operate from A.C. mains, and supply a low voltage regulated D.C. These power supply units perform three essential functions:

(i) They reduce the level of the input voltage and provide isolation from the A.C. mains.

(ii) They provide a means of storing energy.

(iii) They regulate or stabilize the output against variations in either input or load conditions.

Ideally, these functions are performed by a power supply that is small, light-weight, and efficient.

We will briefly consider each of these main functions in more detail. Voltage reduction and isolation from the mains are performed by a transformer. The size, and hence the weight, of the transformer is inversely proportional to the operating frequency (within such design parameters as turns ratio, core material and losses). To keep the size down to a minimum, the transformer should operate at the highest possible frequency.

The reservoir or filter capacitor performs two functions. It (1) stores energy and (2) filters the rectifier output voltage. The energy stored is proportional to $\frac{1}{2} CV^2$. Therefore, the *higher* the voltage level at which the energy can be stored, the smaller need be the size of the capacitor.

Regulation is the third essential function to be considered. Regulation was discussed at some length in Chapters 3 and 4. Both discrete and integrated approaches were presented. They both come into the category of 'linear regulators'. Regulation is obtained by having more power than the load requires and dissipating the excess energy in the form of heat. The physical size is large, due to the heatsinking requirements, and the efficiency is poor. Figure 7.1 shows a block diagram of the linear system.

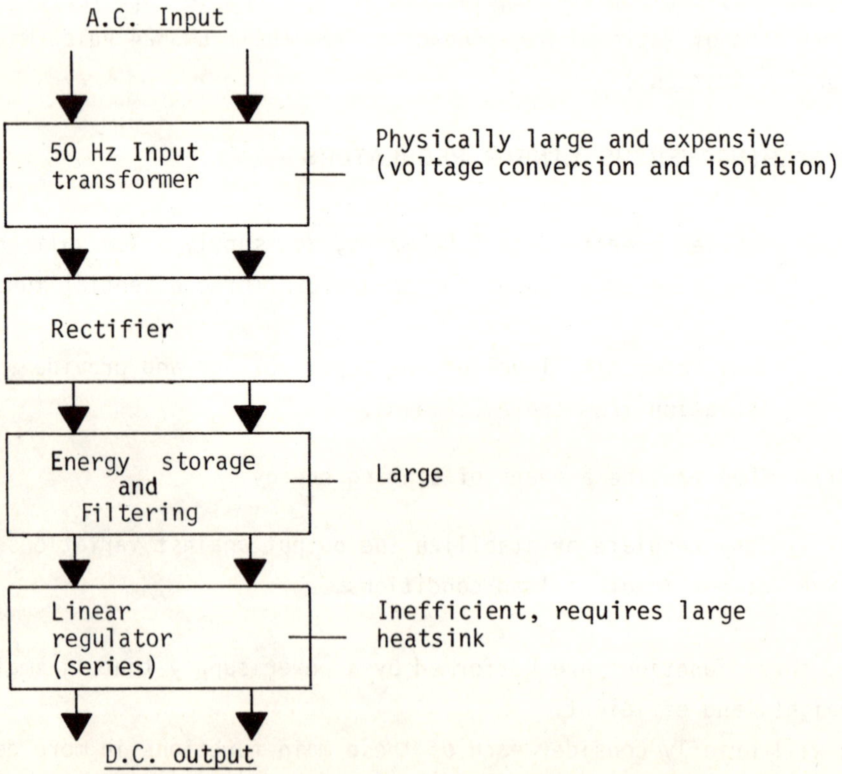

Figure 7.1 Linear regulator system

In contrast, the power dissipated by the regulating device in a switching system is very small. Minimum, if any, heatsinking is required, and because there is negligible power loss, efficiency is high. However, due to the method of controlling the switching device, the average value of the output waveform has to be recovered by rectification and filtering. This results in double rectification, but the power dissipation is nevertheless reasonable, and overall the switching system is smaller and more efficient. Figure 7.2 shows a block diagram of the switching system.

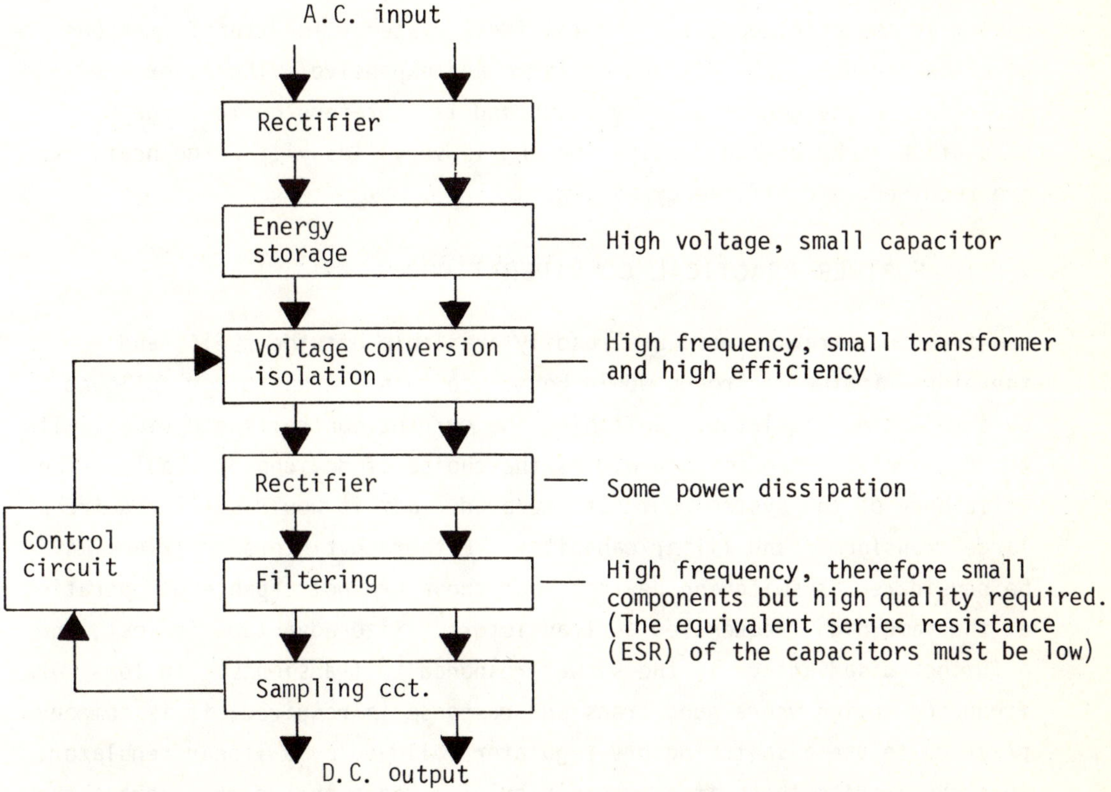

(Caution: this particular system is directly off the A.C. mains, and therefore, all primary circuit components are at mains potential. Use an isolation transformer when testing or servicing this type of power supply.)

Figure 7.2 Switching regulator system

Having discussed the basic requirements for a power supply circuit, we can see that the 'ideal' system would have:

(i) A high frequency transformer

(ii) A high voltage reservoir capacitor

(iii) A switching regulator

If we compare Figure 7.1 and 7.2, it is interesting to note that the linear system is the exact opposite to the ideal system. The transformer operates at a low frequency (50 Hz) and is large and expensive. The reservoir capacitor is charged at a low voltage and is therefore quite large. Regulation is by energy dissipation and is very 'lossy'. Large heatsinks are required, and efficiency is low.

7.2.1 FURTHER PRACTICAL CONSIDERATIONS

Low-frequency transformers are readily available off the shelf, and therefore, another approach would be to use such a transformer followed by a switching regulator. Switching the current-controlling device limits the dissipation stresses and widens the choice of devices available. The efficiency of the system would be improved, but it would still require a large transformer and filter capacitor. Either thyristors or triacs may be considered as switching devices, but these are not capable of operating at the same high frequencies as transistors. Size advantage is lost, and a further disadvantage is the slower response to transients. In low-frequency design where good transient response is required, it is common practice to use a switching pre-regulator followed by a linear regulator. In such circuits the switching pre-regulator keeps the voltage across the linear regulator at a low constant value, irrespective of other circuit variations. This results in the reasonable efficiency and good transient response of a linear regulator.

The choice of the 'type' of output capacitor is also important. No capacitor is entirely free of resistance, but the type of capacitors used in S.M.P.S. must exhibit a low 'equivalent series resistance' (E.S.R.). The E.S.R. is determined by the construction techniques, and such capacitors are more expensive than the conventional type that have a much higher series resistance.

7.2.2 BASIC SWITCHING REGULATOR CONCEPT

The basic operation of a switching regulator can best be understood with the aid of Figure 7.3. Transistor Q_1 is the switching element which is turned on and off by a pulse waveform with a fixed duty cycle (on/off period). The diode D provides a continuous path for the inductor current when the transistor turns off. The voltage waveform on the emitter of the transistor (at Point X) will be as shown in Figure 7.3b. Neglecting any voltage drop across the transistor and the diode, the output of the LC network will be the average value of the switched waveform.

$$V_{out} = V_{in} \times \frac{t_{on}}{T}$$

where T is the total period and t_{on} is the time 'on'. V_{out} is independent of the load current.

It can be seen from the equation above that any changes in V_{in} can be compensated for by varying the duty cycle of the switched waveform. This is what happens in a switching regulator circuit and forms the basis of its operation. The control pulses are referred to as 'pulse width modulated' (P.W.M.).

7.3 BASIC CIRCUITS

As mentioned earlier, the advantages of a switching regulator circuit were its ability to (i) step-down, (ii) step-up, or (iii) invert the input. We will now examine the operation of each of these circuit configurations in detail.

7.3.1 BASIC STEP-DOWN SWITCHING REGULATOR

The basic circuit of a 'step-down' swtiching regulator is shown in Figure 7.4. Transistor Q_1 is the switching element which has the 'on' and 'off' times controlled by a pulse width modulator. When Q_1 is on, power is drawn from V_{in} and supplied to the load. Current passes through the inductor L, and Point V_A is at approximately the potential of V_{in}, i.e. Q_1 is saturated. This positive potential on D_1 cathode reverse biases the diode, and the output capacitor (C_o) is charging. When Q_1 turns off, the magnetic field stored in the inductor collapses. The polarity of the voltage across L

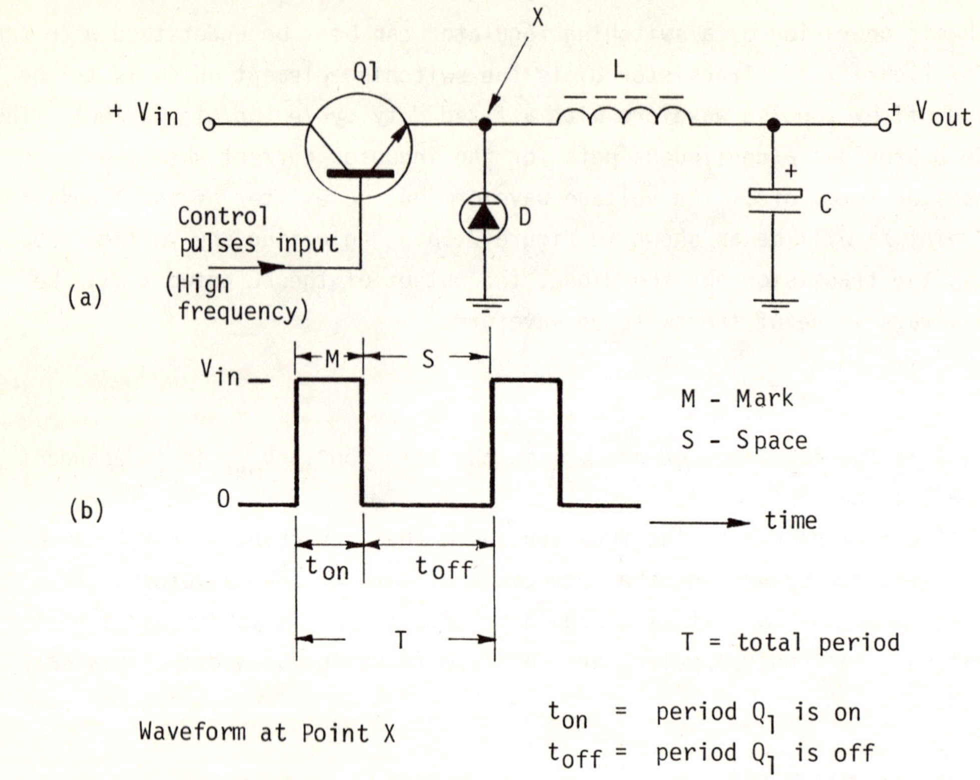

Figure 7.3 Basic circuit and idealized waveform for a switching regulator

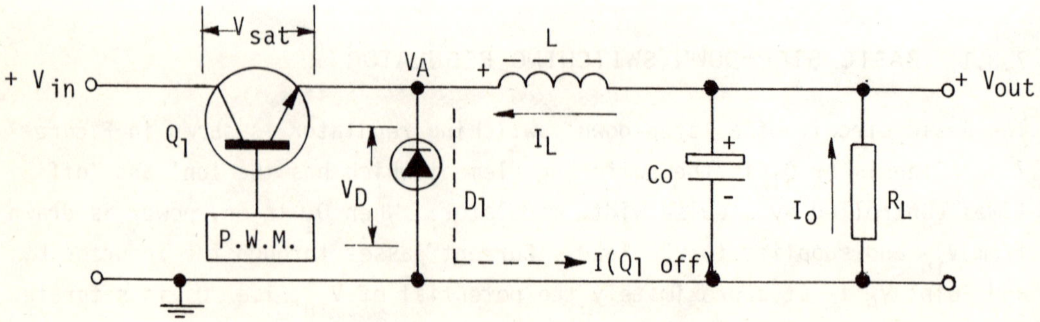

Figure 7.4 Step-down switching regulator (series switched-mode or forward circuit)

reverses, forcing Point V_A negative, and forward biasing the diode (D_1). Current continues through the load and the inductor via the diode.

The control circuit 'samples' the output voltage and automatically increases the 'on' time of the transistor if the output decreases. Likewise, it increases the 'off' time if the output voltage increases. The output voltage is filtered by the LC network and maintained at an almost constant level. As the energy stored in the inductor is recovered during the transistor 'off' time when no current is being taken from the input, the circuit operates at high efficiency.

The current through the inductor is equal to the nominal D.C. load current plus a changing current that results from the changing voltage across the inductor. The peak-to-peak value of the changing current (ΔI_L) is approximately 40 per cent of the output current I_o.

This type of circuit can generate a relatively large amount of noise in the input line due to the rapid switching. The output voltage is approximately equal to $V_{in} - (V_{sat} + V_L)$ and can be determined by the formula:

$$V_{out} = V_{in} \times \frac{t_{on}}{T}$$

where T is the total period for one cycle, and t_{on} is the period Q_1 is on. The voltage and current waveforms are shown in Figure 7.5.

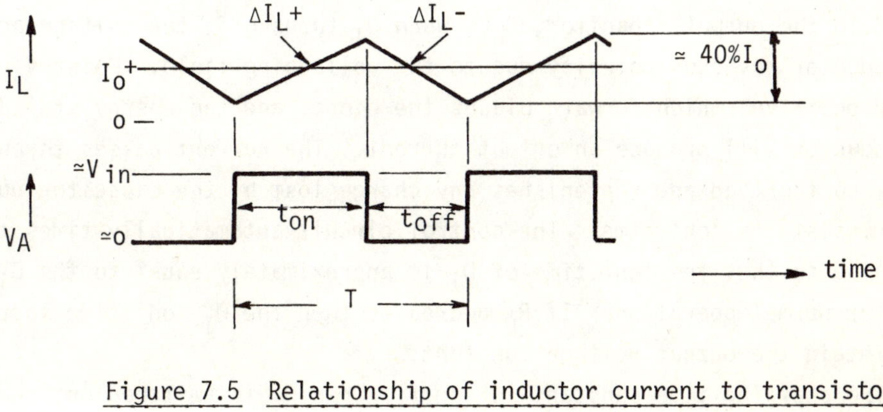

Figure 7.5 Relationship of inductor current to transistor on/off time in a step-down circuit

Efficiency approaches the ideal 100 per cent when the transistor and diode voltage drops are small. The output ripple is dependent on the duty cycle, the frequency of switching, and the size of the output capacitor C_o.

The larger the capacitor, the lower the ripple amplitude.

7.3.2 BASIC STEP-UP SWITCHING REGULATOR

The basic circuit of a 'step-up' switching regulator is shown in Figure 7.6. Assume C_o is charged. Transistor Q_1 is the switching element and is used

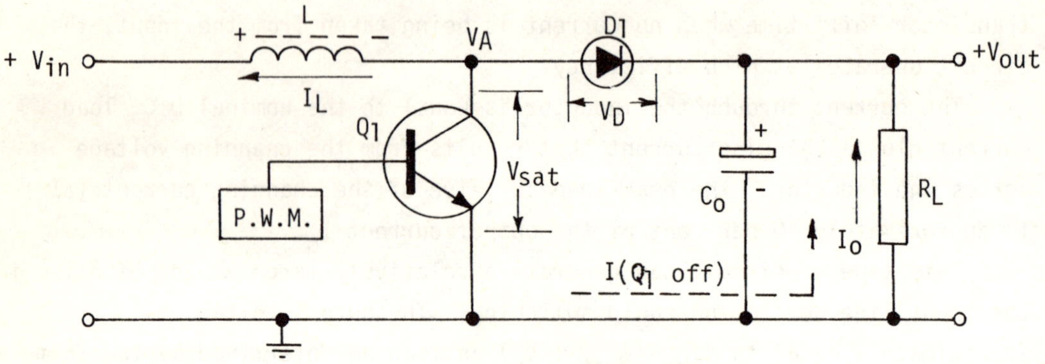

Figure 7.6 Step-up switching regulator

to apply the input voltage across the inductor during the 'on' time. When Q_1 is on, energy is drawn from V_{in} and stored in the inductor. The diode is reverse biased because Point V_A is forced less positive than V_{out}, and the output current is supplied to the load from the charge previously stored in the output capacitor, C_o. When Q_1 turns off, the voltage across the inductor reverses polarity due to the collapsing field. Point V_A is forced positive, which forward biases the diode, and the energy stored in the inductor will produce an output current. The current passes through L and D_1 to the load and replenishes any charge lost by the capacitor during the transistor's 'on' time. The control circuit automatically times the switching so that the 'on' time of D_1 is approximately equal to the Q_1 'on' time for normal operation. If R_L decreases then the Q_1 'on' time increases to maintain the output voltage constant.

The charging time constant is short. When Q_1 is switched on, the energy is stored in L very quickly,

$$T.C = \frac{L}{R \text{ of } Q_1}$$

When Q_1 turns off, the voltage developed across L is in series with V_{in}, and the output capacitor charges towards the sum of the two voltages ($V_{in} + V_L$): hence a stepped-up voltage at the output. In this type of circuit, the input current must exceed the output current to enable energy transfer, and so step up the output voltage.

This circuit has the advantage that very little noise is generated in the input line. The inductor isolates the input from the switching transistor and helps to smooth out switching noise. Also, current is present in the load circuit during both the 'on' and 'off' states of the transistor, again reducing the switching noise. Unfortunately, this makes output filtering more difficult. The voltage and current waveforms are shown in Figure 7.7. The relationship between input voltage, output voltage, and duty cycle can be seen from the equation:

$$V_{out} \cong V_{in} \left(1 + \frac{t_{on}}{t_{off}}\right)$$

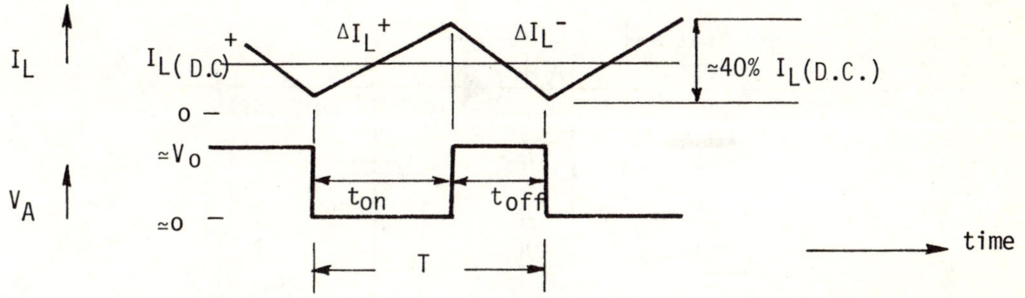

Figure 7.7 Relationship of inductor current to transistor on/off time in a step-up circuit

When the voltage drops across the transistor and the diode are small compared to V_{in} and V_{out}, the efficiency approaches 100 per cent. Ripple can be reduced by using a large value of output capacitor.

7.3.3 BASIC INVERTING SWITCHING REGULATOR

The basic circuit of an 'inverting' switching regulator is shown in Figure 7.8. This circuit is also known as a shunt switched-mode circuit or 'flyback' type circuit. It provides a negative output voltage from a positive input voltage. The magnitude of the output voltage may be either larger or smaller than the input, depending on circuit design. Transistor Q_1 is again the switching element, and applies the input voltage across the inductor during the 'on' time. When Q_1 is on, energy is drawn from the supply and stored in the inductor. When Q_1 turns off, the voltage across the inductor reverses due to the collapsing field, and the diode is forward biased. Point V_A is thus forced negative. The energy stored in the inductor produces an output current in the load in the opposite direction and charges the output capacitor C_o.

The control circuit automatically adjust the transistor's 'on' and 'off' time to produce an almost constant output voltage of the opposite polarity. In this type of circuit, both the input and output circuits are quite noisy at the switching frequency. Normal electrolytics are insufficient as filtering due to their inductance at high frequencies, and therefore a small value capacitor is required to shunt the high frequencies.

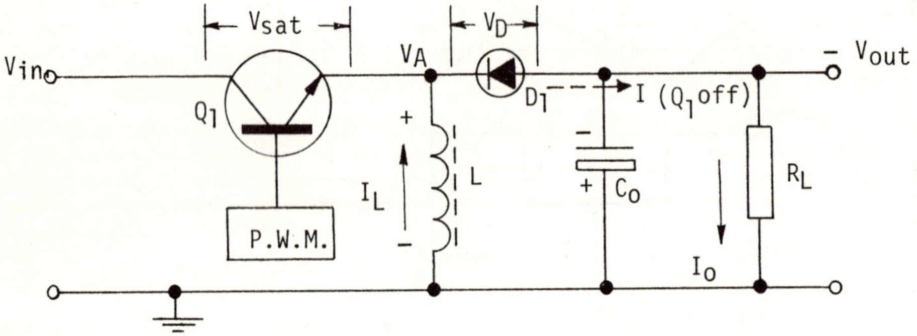

Figure 7.8 Inverting switching regulator

Efficiency is high if V_{sat} is low during the 'on' time and V_D is small when the energy is transferred to the load. The size of the output capacitor will again determine the amplitude of the ripple.

The shunt switched-mode circuit is the only configuration that is suitable for mains isolation. Figure 7.9 shows the circuit modifications and waveforms to achieve mains isolation. The inductor is replaced by a transformer which isolates the primary and secondary circuit from each other. A turns ratio is selected which will increase or decrease the output voltage; also, multiple secondaries can be used to derive more than one output voltage. This configuration is very popular in colour television switched-mode power supply units.

When the transistor is on, the D.C. input voltage is fed to the transformer primary, and a linear rising current results. The time constant for the 'on' time is short and is determined basically by the resistance of the transformer primary winding. When the transistor switches off, the primary current falls to zero, and the voltages across the transformer windings change polarity (signs shown inside the circles in Figure 7.9a).

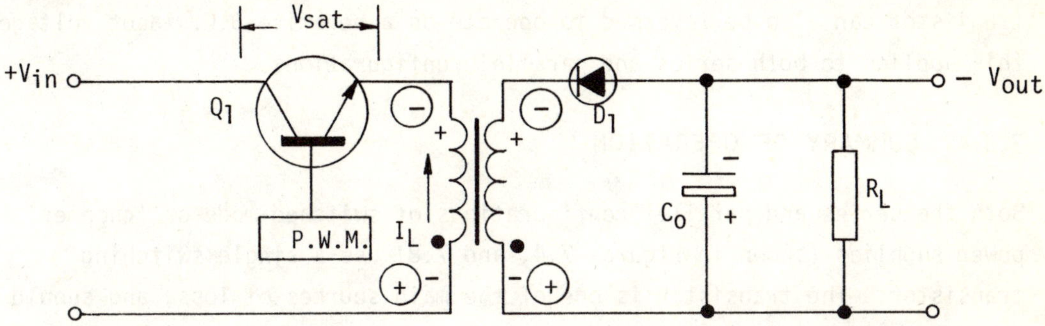

a) Circuit

(Note: With the isolated type of switching transformer, care must be taken with the design so that all the primary turns are 'mutually coupled' to the secondary turns, i.e. no 'leakage inductance', as this puts extra stress on the chopper transistor at switch off.)

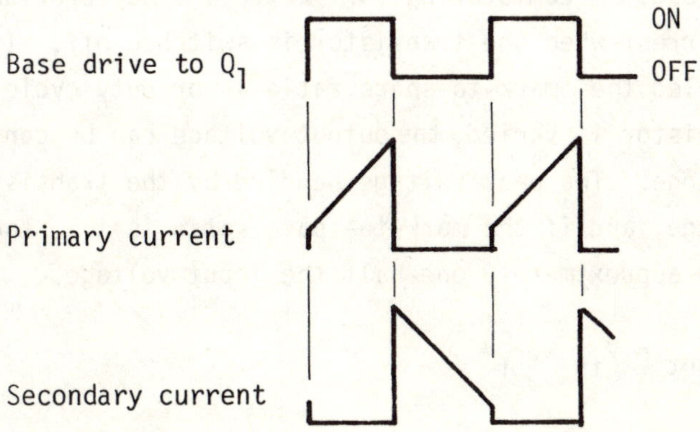

b) Waveforms

Figure 7.9 Isolated parallel switching regulator and 'basic' waveforms

Diode D_1 in the secondary circuit is now forward biased, and current passes to the output capacitor and load. The secondary current falls linearly with time, dissipating the stored magnetic energy. The secondary time constant is long compared to the primary time constant. This means that the secondary current will not fall to zero before the transistor conducts again. The circuit operation then repeats itself.

By choosing the phase relationship of the transformer windings and diode polarity, the output can be inverted or non-inverted. The switching transistor can also be reversed to operate on a negative D.C. input voltage; this applies to both series and parallel configurations.

7.3.4 SUMMARY OF OPERATION

Both the series and parallel configurations of switched-mode or 'chopper' power supplies (shown in Figures 7.4, and 7.8) use a single switching transistor. The transistor is one of the main sources of loss, and should therefore be capable of switching at high frequencies. This means that the rise and fall times should be short. The diode for the reverse recovery of the inductor's energy should also be as fast as possible.

In the series configuration, a D.C. input voltage is obtained from the mains and chopped by the transistor, which is driven by a square or rectangular wave. The 'chopped' voltage is smoothed by a L.C. filter and passed to the load. A commutating (or 'catch') diode provides a path for the inductor current when the transistor is switched off. If the on/off time (often called the 'mark-to-space ratio') or duty cycle of the switching transistor is varied, the output voltage can be controlled over quite a wide range. The peak voltage handled by the transistor is equal to the input voltage, and if the mark-to-space ratio is 1:1, then the output voltage will be approximately one-half the input voltage.

$$V_{out} = V_{in} \times \frac{t_{on}}{T}$$

If V_{in} = 24V, t_{on} = 1.0 microseconds, and T = 2.0 microseconds (i.e. mark-to-space = 1:1), then:

$$V_{out} = \frac{24}{1} \times \frac{1}{2}$$

$$= 12 \text{ volts}$$

In the shunt configuration, energy from the rectified A.C. supply is stored in the inductor when the transistor conducts. When the transistor is off, energy is transferred via the diode to the load. The output voltage will be the opposite polarity to the input unless transformer coupling is used. Again, the output voltage can be controlled by varying the mark-to-space ratio of the switching circuit. If the mark-to-space ratio is 1:1, then the output voltage will be equal to the input voltage, because energy is stored for one half-cycle, and released to the load on the next half-cycle. As mentioned in Chapter 3, shunt type regulator circuits are more heavily stressed than series regulator circuits. This also applies to switched-mode supplies, but they have the advantage of providing mains isolation if the inductor is replaced by a transformer.

Finally, the choice exists between forward and flyback configurations. A forward S.M.P.S. circuit transfers energy to the inductor and the load when the transistor is on. A flyback S.M.P.S. circuit stores energy in the inductor when the transistor is on and transfers it to the load when the transistor is off. Figure 7.10 shows a graph of the general voltage-power limits within which each type of circuit is more efficient.

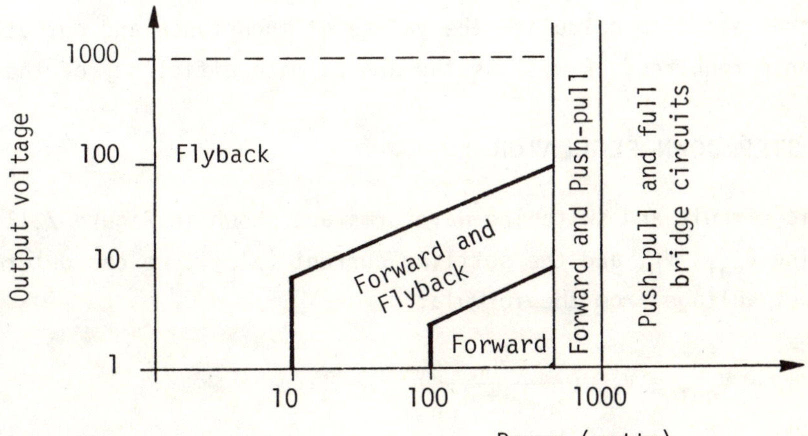

Figure 7.10 Voltage - power recommendations for different circuit configurations

In the series step-down switched-mode circuit, the inductor or 'choke' is connected in series with the load, and energy is transferred to the load while the transistor is 'on'. In the shunt switched-mode circuit, the inductor is in parallel with the load. Energy is stored in the choke while the transistor is on, and delivered to the load when the transistor if off. Advantages of the series type circuit include (1) switching that is limited to the transistors 'on' period, (2) a peak collector-emitter voltage (V_{CE}) that is no greater than the D.C. input voltage, (3) a smaller inductor than the parallel circuit, and (4) a lower level of output ripple current. The disadvantages include impossibility of obtaining mains isolation and a distinct possibility that the full input voltage will appear across the load if the switching transistor shorts.

7.4 DESIGN INFORMATION

(Note: Basic design information in this section is reproduced by kind permission of National Semiconductor)

When building a switched-mode power supply, careful consideration must be given to the values of various circuit components. Different circuit configurations require slightly different components. This section will enable the reader to calculate the values of inductance and output capacitance required, as well as the approximate efficiency of the circuit.

7.4.1 STEP-DOWN REGULATOR

The basic circuit and switching waveforms are shown in Figure 7.11. Neglecting V_{sat}, V_D, and the settling current ($\Delta I_L \pm$), we can determine the output voltage from the formula:

$$V_{out} = V_{in} \left(\frac{t_{on}}{t_{off} + t_{on}}\right)$$

$$V_{out} = V_{in} \left(\frac{t_{on}}{T}\right)$$

where T = total period

t_{on} = period Q_1 is on

t_{off} = period Q_1 is off

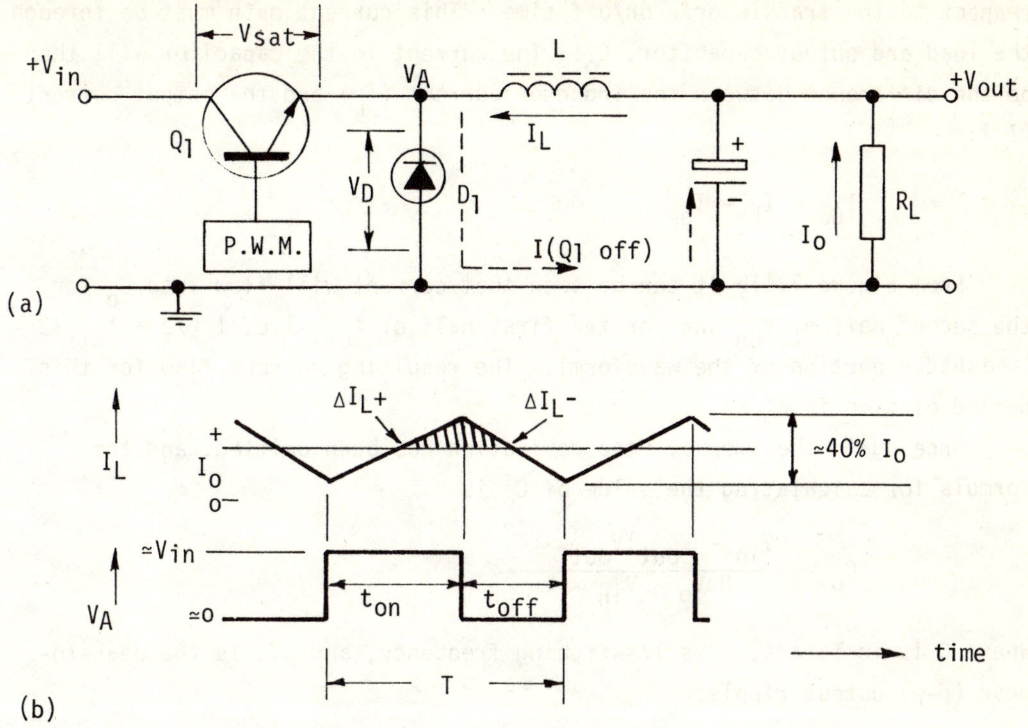

Figure 7.11 Basic circuit and switching waveforms

The formula shows the relationship between V_{in}, V_{out}, and duty cycle. Transistor Q_1 only conducts during the 'on' time.

$$\text{Efficiency} = \frac{V_{out}}{V_{out} + 1} \text{ for } V_{sat} = V_D = 1 \text{ volt.}$$

The maximum efficiency will decrease further due to switching losses for Q_1. *The switching transistor should therefore have very fast rise and fall times for all circuit configurations.*

Calculating Inductor L. The mathematics is fairly complex, and for this reason the step by step derivation has been omitted.

$$L = \frac{2.5 \times V_{out}(V_{in} - V_{out})}{I_{out} \times V_{in} \times f}$$

where L is in henrys and f is the switching frequency in hertz.

Calculating Capacitor, C_o. Figure 7.11b shows the inductor's current with respect to the transistor's on/off time. This current path must be through the load and output capacitor, C_o. The current in the capacitor will then be the difference between the inductor current (I_L) and the output current (I_o):

$$I_{Co} = I_L - I_o$$

From Figure 7.11b it can be seen that current will flow into C_o for the second half of t_{on} and for the first half of t_{off} i.e. $t_{on}/2 + t_{off}/2$ - (the shaded portion of the waveform). The resulting current flow for this period of time is $\Delta I_L/4$.

Once again the step by step derivation has been omitted, and the formula for calculating the value of C_o is

$$C_o = \frac{(V_{in} - V_{out})V_{out} \times T^2}{8\Delta V_o \times V_{in} \times L}$$

Where C is in farads, T is 1/switching frequency, and ΔV_o is the peak-to-peak (p-p) output ripple.

7.4.2 STEP-UP REGULATOR

The basic circuit and switching waveforms for a step-up regulator are shown in Figure 7.12a and b. Neglecting V_{sat} and V_D, and noting that $\Delta I_L+ = \Delta I_L-$, then:

$$V_{out} = V_{in}(1 + t_{on}/t_{off})$$

To calculate the input current I_{in} (D.C.), which equals I_L (D.C.), 100 per cent efficiency must first be assumed.

$$P_{in} = I_{in}\ (D.C.) \times V_{in}$$

$$P_{out} = I_{out}V_{out} = I_{out} \times V_{in}(1 + t_{on}/t_{off})$$

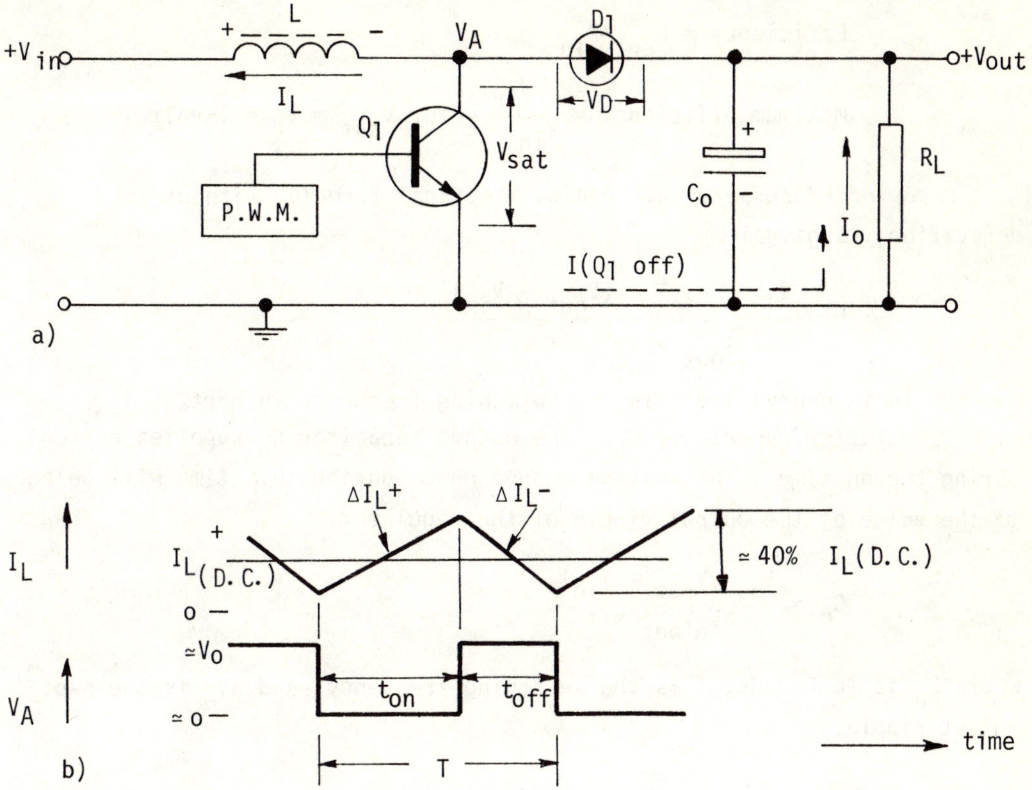

Figure 7.12 Step-up regulator

If efficiency is 100 per cent, then:

$$P_{out} = P_{in}$$

$$I_{out} V_{in} (1 + t_{on}/t_{off}) = I_{in}\ (D.C.)\ V_{in}$$

Therefore

$$I_{in}\ (D.C.) = I_{out}(1 + t_{on}/t_{off})$$

This shows that the input, or inductor current, is larger than the output, as stated earlier, by the factor $(1 + t_{on}/t_{off})$. This is the same relationship as V_{in} and V_{out}.

$$V_{out} = V_{in}\ (1 + t_{on}/t_{off})$$

$$\text{Efficiency} = P_{out}/P_{in}$$

$$\text{Maximum efficiency} = \frac{V_{in}}{V_{in} + 1} \text{ for } V_{sat} = V_D = 1 \text{ volt}$$

Calculating Inductor L. Again, the final formula, without its derivation, is given:

$$L = \frac{2.5 \times V_{in}^{2} \times (V_{out} - V_{in})}{I_{out} \times V_{o}^{2} \times f}$$

where L is in henrys and f is the switching frequency in hertz.

Calculating Capacitor, C_o. The output capacitor C_o supplies current during the on time. The voltage change on C_o during this time will be equal to the value of the output ripple of the regulator.

$$C_o = \frac{I_{out}(V_{out} - V_{in})}{\Delta V_o V_{out} \times f}$$

where C_o is in Farads, f is the switching frequency, and ΔV_o is the p-p output ripple.

7.4.3 INVERTING REGULATOR

The basic circuit and switching waveforms of an inverting regulator, are shown in Figure 7.13. Energy builds up in the inductor during the 'on' time, and is delivered to the load during the 'off' time. The circuit is therefore a flyback type. The values of L and C_o are calculated from the following formulas.

$$L = \frac{2.5 \times V_{in} \times V_{out}}{(V_{out} + V_{in})I_{out} \times f}\ .$$

where L is in henrys and f is the switching frequency in hertz.

$$C_o = \frac{I_{out}V_{out}}{\Delta V_o \times f \times (V_{out} + V_{in})}$$

where C_o is in Farads, f is the switching frequency in hertz, and ΔV_o is the p-p output ripple. The output voltage is dependent on V_{in}, the duty cycle,

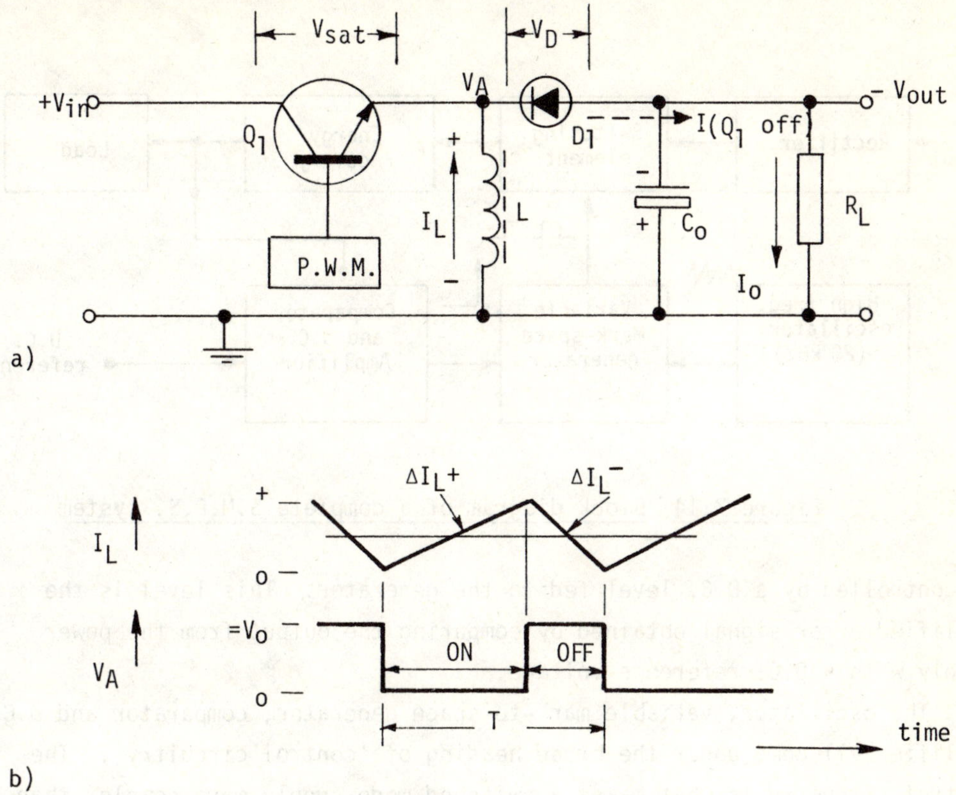

Figure 7.13 Basic circuit and switching waveforms

and the inductor. Neglecting V_{sat} and V_D,

$$V_{out} = V_{in}(t_{on}/t_{off})$$

7.5 CONTROL CIRCUITRY

The basic operation of S.M.P.S. configurations has been discussed in some detail. All of the circuits use a form of automatic control to sample the output voltage and adjust the mark-to-space ratio of the control circuit accordingly. An overall block diagram of the switched-mode system is shown in Figure 7.14.

The switching element is operated at a fixed high frequency (approximately 20 kHz.) A variable mark-to-space ratio generated accepts this 20 kHz, and drives the switching element. The mark-to-space ratio

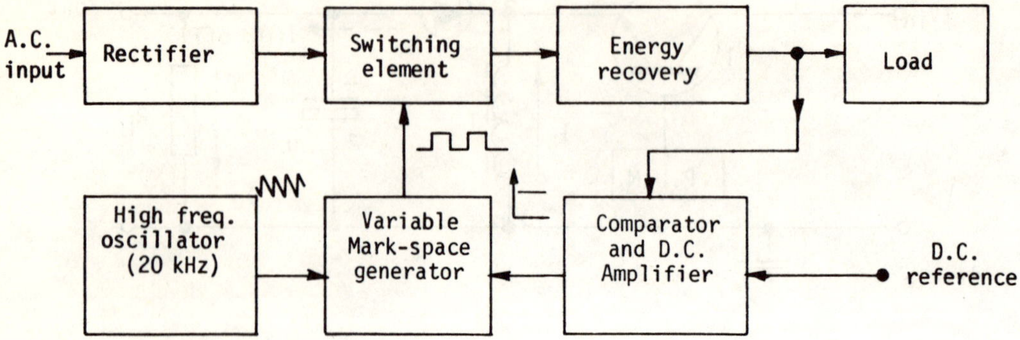

Figure 7.14 Block diagram of a complete S.M.P.S. system

is controlled by a D.C. level fed to the generator. This level is the amplified error signal obtained by comparing the output from the power supply with a D.C. reference voltage.

The oscillator, variable mark-to-space generator, comparator and D.C. amplifier all come under the broad heading of 'control circuitry'. The control circuitry is what makes a switched-mode supply more complex than a linear supply, and it can be found in both discrete or integrated forms. With discrete control circuits, the oscillator is usually a transistor multivibrator or unijunction transistor. The comparator/D.C. amplifier is a transistor amplifier or differential pair, and the output circuit has series current limiting with over voltage protection. There are many components in this type of circuit, but new technology has incorporated the whole lot in one integrated circuit. This has made switched-mode power supplies a very viable proposition by reducing their complexity. Many companies are producing switched-mode control chips, and Figure 7.15 lists the characteristics of a few of them.

Two basic 'block diagrams' of control chips are shown in Figure 7.16 and 7.17.

It is impossible to cover every variation and make of switched-mode controller I.C. in the text. Further information can be obtained from the following manufacturers or their representatives: Exar, Fairchild, Ferranti, Hitachi, Motorola, National Semiconductor, Plessey, and Texas Instruments.

Type	Output configuration	Accessible E & C outputs	On-chip error op-amp.	On-chip overvoltage or current protection	Current limited output stage
494	push pull	✓	✓	✓	X
3520	push pull	collectors only	X	X	X
5560	single ended	✓	✓	✓	✓
1524	push pull	✓	✓	✓	X

Figure 7.15 Comparison of switched-mode controller characteristics

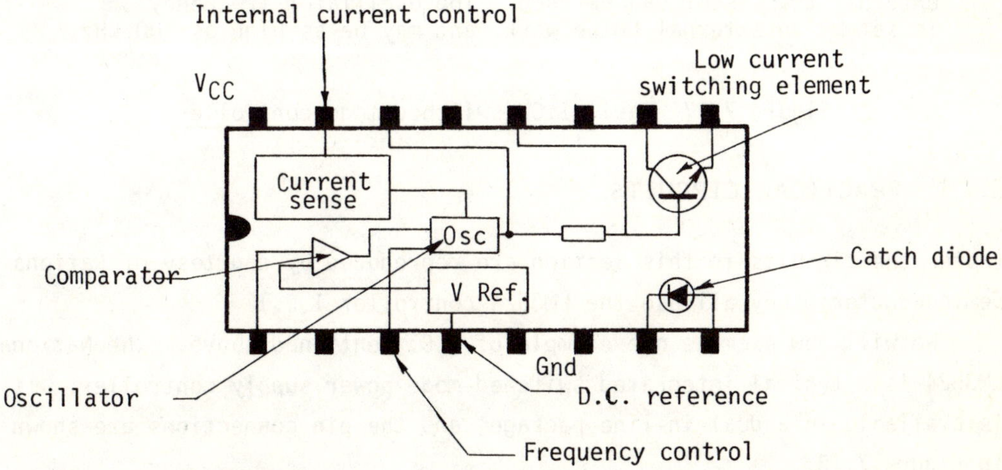

I.C. containing basic control circuitry and a single-ended switching element. Approximate output current is 200 mA, and if higher output current is required, an external transistor is used. Even the catch diode is incorporated inside this I.C., and very few external components are required.

Figure 7.16 Basic I.C. switched-mode controller

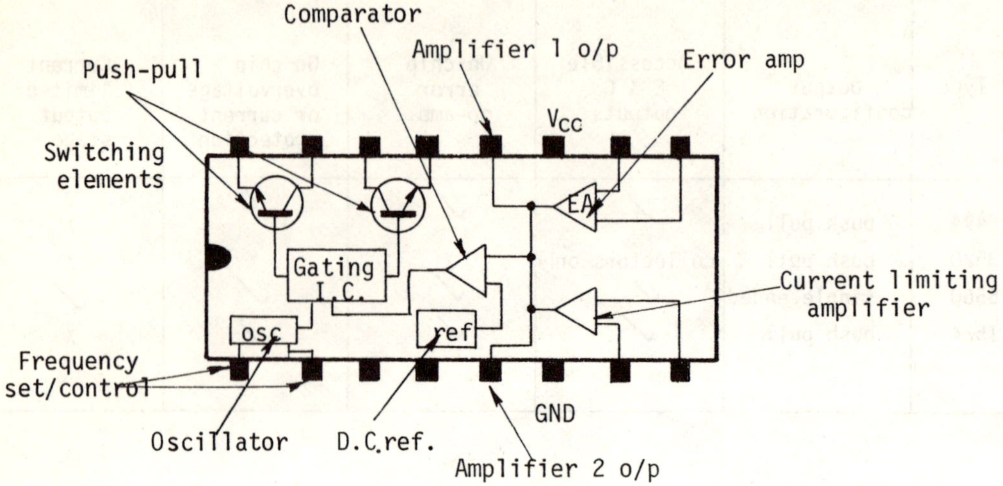

I.C. containing the control circuitry. The switching output is push-pull, and if higher output current is required an external transistor can be used. The oscillator frequency is set by an external CR network, and may be as high as 100 kHz.

Figure 7.17 Basic I.C. switched-mode controller

7.5.1 PRACTICAL CIRCUITS

(Note: the circuits in this section are reproduced by courtesy of National Semiconductor; they all use the LM3524 controller I.C.)

We will now examine one example of I.C. mentioned above. The National LM3524 is a typical integrated switched-mode power supply controller. It is available in a dual-in-line package, and the pin connections are shown in Figure 7.18.

The LM3524 integrated regulating pulse-width modulator contains all of the circuitry that is required to control a switched-mode power supply. The device has a control amplifier, an oscillator, a pulse-width modulator, a phase-splitting flip-flop, dual-output transistors, and a current-limiting shutdown circuit. It also contains a built-in 5 volt voltage regulator that has a current capability of 50 mA (at Pin 16).

The oscillator frequency is set by two external components: a resistor R_T and a capacitor C_T (Pins 6 and 7). The output of the oscillator (Pin 3)

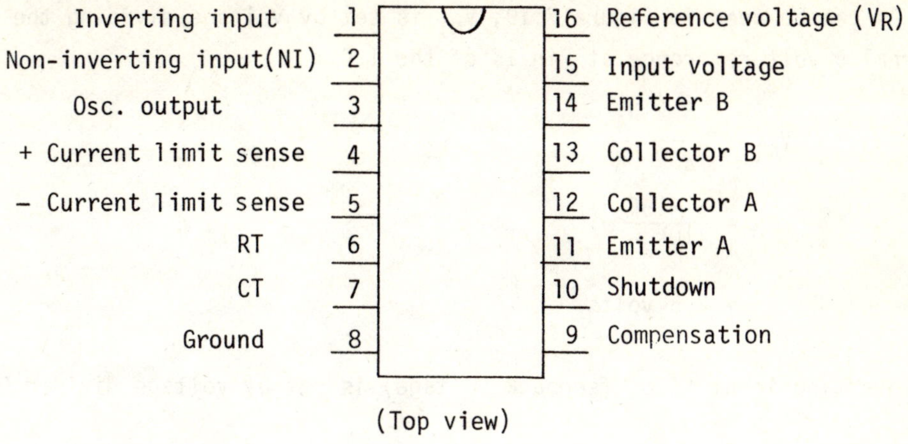

Figure 7.18 Pin connections of the LM3524 S.M.P.S. controller

triggers an internal flip-flop, which provides the pulse-width modulated signal to the output transistors.

The output transistors are NPN types with a maximum current capability of 100 mA. The bases are driven internally, 180 degrees out of phase, but the collectors and emitters are accessible at Pins 12 and 13, and 11 and 14 respectively.

The error amplifier has a differential input through Pins 1 and 2, and the gain is set by a feedback voltage divider taken from the output. The error amplifier can be overridden by applying a D.C. voltage to Pin 9 of the I.C.; this will reduce the duty cycle of the pulse-width modulator.

Likewise, the current-limiting circuit takes control of the error amplifier and changes the pulse duty cycle. A resistor is used to 'sense' the output current, and the voltage developed is applied between the $\pm C_L$ pins (4 and 5) of the I.C. A voltage of 200 mV will reduce the duty cycle by 25 per cent.

A complete step-down switching regulator circuit schematic is shown in Figure 7.19. Transistors Q_1 and Q_2 have been included to boost the output current to 1 ampere - normally 80 mA is available from the I.C. internal transistors. The output voltage is determined by the formula

$$V_{out} = V_{NI} \left(1 + \frac{R_1}{R_2}\right)$$

where V_{NI} is the voltage at the error-amplifier's non-inverting input.

As can be seen in Figure 7.19, V_{NI} is set by voltage dividing the internal 5 volt reference at Pin 16 of the I.C.:

$$V_{NI} = \frac{5 \times R_4}{R_4 + R_5}$$

$$= \frac{5 \times 5k}{10k}$$

$$= 2.5 \text{ volts}$$

The inverting input (i.e. feedback voltage) is set by voltage divider R_1 and R_2.

$$= \frac{V_{out} \times R_2}{R_1 + R_2}$$

$$= \frac{5 \times 5k}{10k}$$

$$= 2.5 \text{ volts.}$$

Resistor R_6 and capacitor C_1 are connected to Pins 6 and 7 to set the oscillator frequency at approximately 20 kHz. Transistor Q_2 is controlled by the I.C. and this in turn controls the conduction of the series pass element Q_1 (R_9 and R_{10} provide the initial forward bias required for Q_1 and Q_2). Inductor L stores the energy while Q_1 is conducting. Diode D is the catch diode for the output current path when Q_1 is off, and the magnetic field around L collapses. Capacitors C_3 and C_4 provide high- and low-frequency filtering at the input. Capacitor C_6 is the output storage capacitor, and C_5 provides high frequency decoupling. Resistor R_3, connected between Pins 4 and 5, sets the current limiting.

$$\frac{V_S}{R_3} = \frac{200 \text{ mV}}{R_3} = \frac{200 \text{ mV}}{0.15} = 1.3 \text{ A}$$

(Output duty cycle = 25 per cent, when V_S = 200 mV.)

This circuit should display the following characteristics: (i) output voltage-5 volt, (ii) operating frequency-20 kHz, (iii) short circuit current limit-1.3 A, (iv) output ripple 10 mV peak to peak, and (v) an efficiency of 80 per cent.

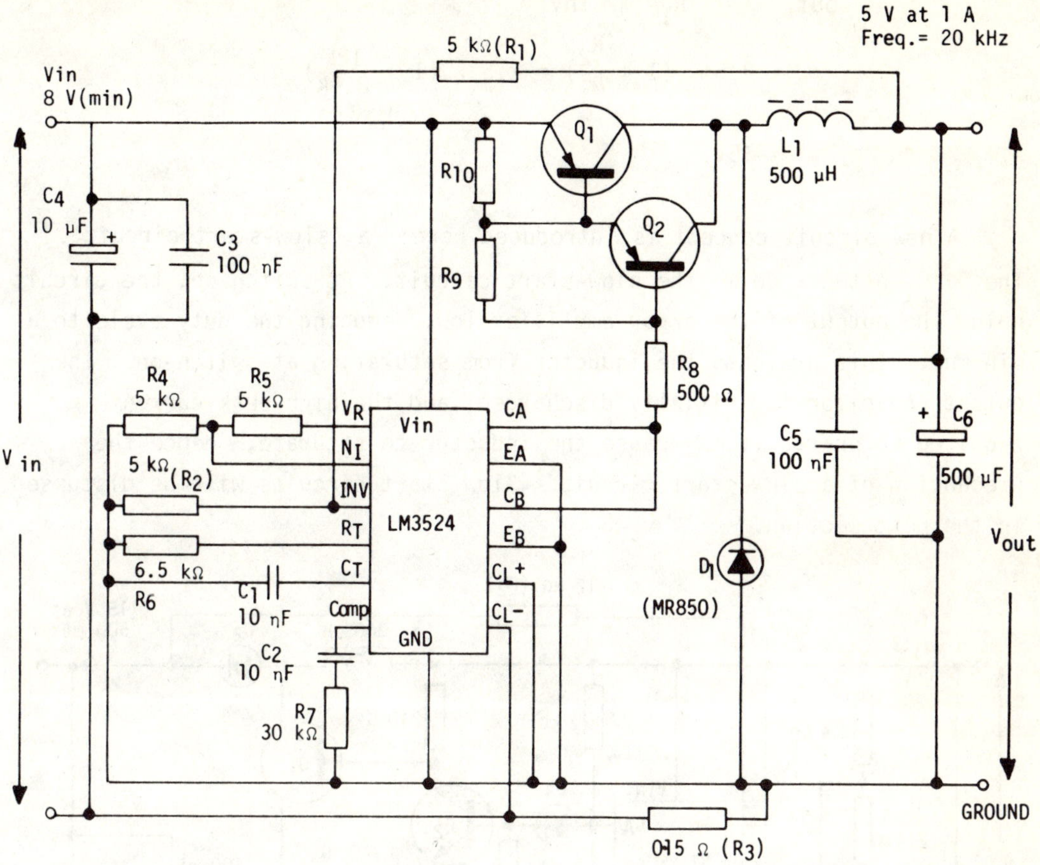

Q_1 = BD 344 (Mounted on a small heatsink)

Q_2 = 2N5023

L_1 = Approx. 40 turns No. 22 wire on a Ferrox cube K300502 core.

Figure 7.19 A 5 V, 1 A step-down switching regulator

A step-up switching regulator is our next example, again using the National LM3524 controller. The circuit schematic is shown in Figure 7.20, and this circuit steps an unregulated 5 volt input up to 15 volts at a current of 500 mA. This time the input voltage is divided into two to bias the error amplifier's non-inverting input, and the output voltage is:

$$V_{out} = (1 + \frac{R_2}{R_1}) \times V_{Inv}$$

$$= 2.5 \times (1 + \frac{R_2}{R_1}) = 2.5 \times (1 + \frac{12k}{2.4k})$$

$$= 15V.$$

A new circuit concept is introduced here: a 'slow-start circuit'. The D_1-C_1 network forms the slow-start circuit. At switch-on, the circuit holds the output of the error amplifier low, reducing the duty cycle to a minimum. This prevents the inductor from saturating at switch-on. The output capacitor is initially discharged, and the high-peak current required to charge it may cause the inductor to saturate; hence the precaution of a slow-start circuit. Slow start circuits will be discussed in the next section.

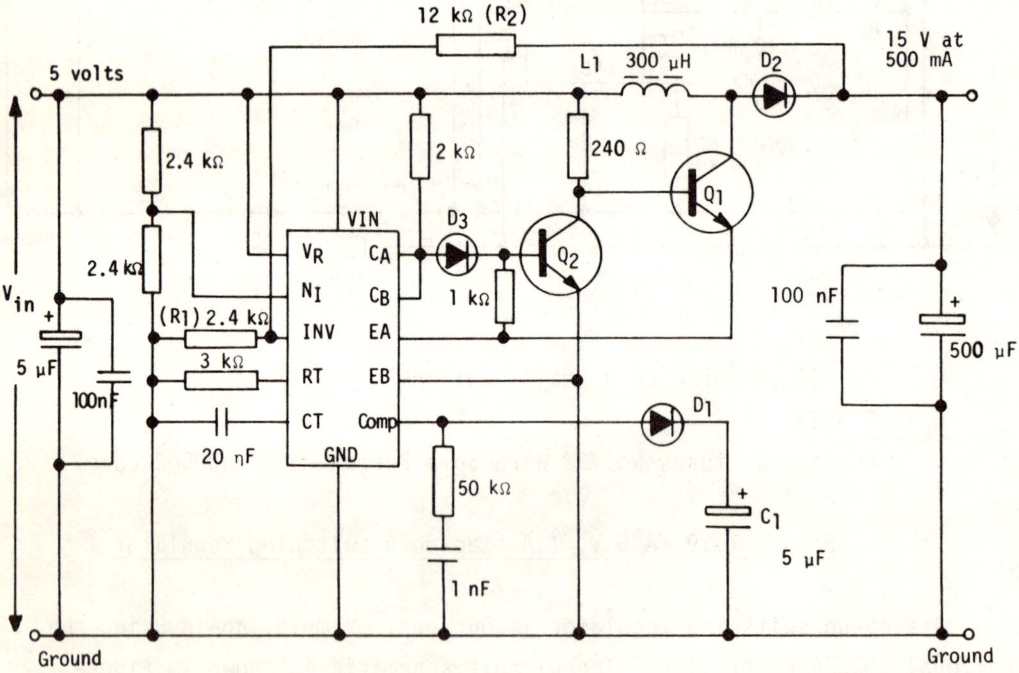

D_1 & D_3 = 1N914 D_2 = MR850
Q_1 = BD345 Q_2 = 2N2219
L_1 = 25 turns of No.24 wire on a Ferrox cube K300502

Figure 7.20 A 15 V, 500 mA step-up switching regulator

7.6 PROTECTION CIRCUITS

The switch-on surge in switched-mode power supplies is usually contained to the first couple of charging-current pulses. When the circuit is first switched on, the large value output capacitor has to be charged, and this causes the output voltage to be low. The comparator in the control circuit sees this as a drop in output voltage and therefore increases the duty cycle to restore the output level. The high current that is present for the initial turn-on period may saturate the inductor or damage the switching element itself. The sudden increase in voltage could also overload the device, blow the fuse, or damage the capacitor. It is desirable that the output voltage should rise gradually to the normal working level over a period of a couple of seconds. The circuit that controls this gradual rise is called a 'soft start' or 'slow turn-on' circuit.

7.6.1 SOFT-START CIRCUITS

The basic concept of all soft-start circuits is to hold the output of the error amplifier low, thereby keeping the duty cycle to a minimum. The duty cycle of the switching transistors, or element, is then allowed to increase gradually to the normal operating point during the circuit's power-up time. This function is particularly easy to implement with integrated circuit controllers. Figure 7.21 shows a typical circuit arrangement.

During power up, the voltage at Pin 3 rises exponentially from zero volts towards V_{CC}, with a time constant of R_1C_1. This allows a gradual increase in the duty cycle. The transistor Q_1 (inside the dotted area) can further extend the soft-start capability to provide an inhibit control, or a soft-start reset. When an inhibit command is received, Q_1 will shunt capacitor C_1, hence discharging it and reducing the voltages at Pins 3, 12 and 5. Diode D_2 allows capacitor C_1 to reset (discharge) when the power is turned off.

Some integrated circuits have the soft-start feature incorporated in them. They only require an external capacitor for the time constant and a discharge resistor to reset the soft start when the power is turned off.

Another approach is to control the conduction of a transistor in a similar manner. The circuit in Figure 7.22 shows this type of operation. The slow-start circuit is connected to the emitter of transistor Q_1, making it conduct heavily to reduce the trigger pulse level and to allow the S.M.P.S. output voltage to rise gradually. Transistor Q_2 is connected in

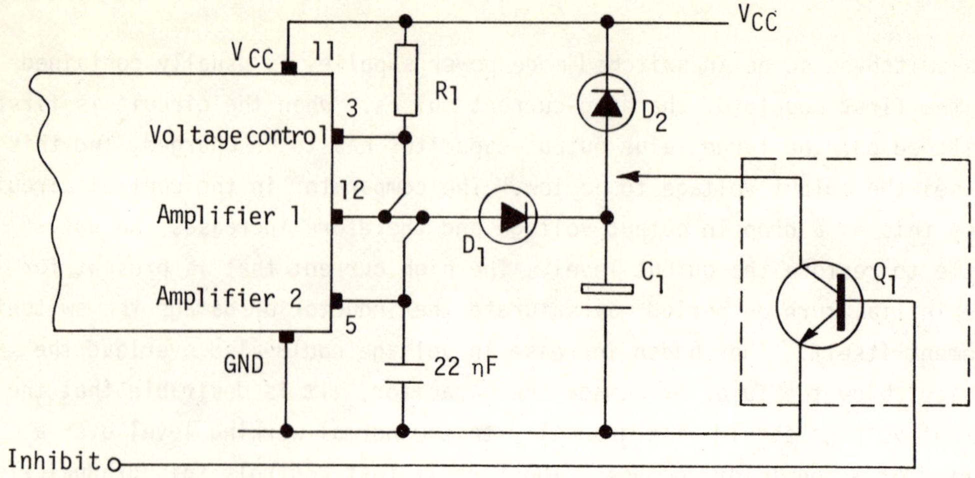

(This diagram features an expanded area of the I.C. in Figure 7.16b)

Figure 7.21 Slow-start circuit

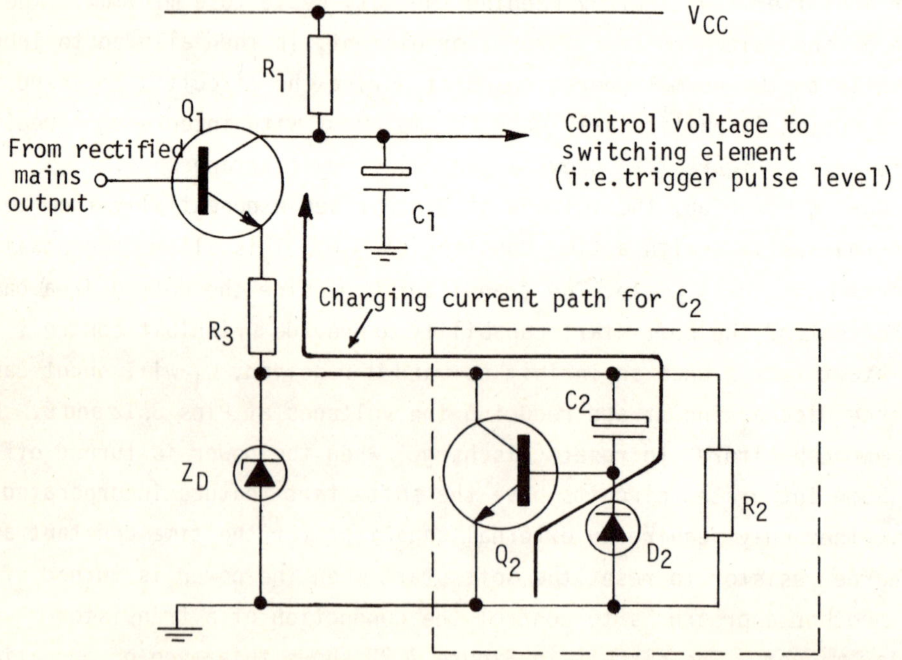

Figure 7.22 Transistor slow-start circuit (inside dotted lines)

parallel with the zener diode (Z_D) in Q_1 emitter circuit. This transistor, (Q_2) is normally non-conducting because there is no base bias voltage. At switch-on, C_2 is discharged, and the collector voltage of Q_2 is applied to the base of Q_2 via C_2. Transistor Q_2 conducts, and C_2 begins to charge through Q_2 base emitter junction. With Q_2 conducting, the zener diode is bypassed, causing the emitter voltage of Q_1 to decrease. Transistor Q_1 has a sharp increase in V_{BE} and conducts heavily. The collector voltage of Q_1 decreases, and this ensures that the voltage across capacitor C_1 takes longer to reach the required voltage level. As capacitor C_2 charges, the base emitter voltage of Q_2 decreases, turning Q_2 slowly off and allowing the zener diode to work again. When C_2 is charged, current ceases to flow in Q_2 base emitter; Q_2 is 'off', reducing the emitter current in Q_1. This allows the collector voltage of Q_1 to rise and charge capacitor C_1 to the voltage required to trigger the switching element. When the equipment is switched off, C_2 discharges through resistor R_2 and the diode D_2. It is thus ready for the next soft start.

7.6.2 CURRENT LIMITING

There are several simple methods to ensure that electronic equipment is protected against excessive current. Some of these include: anti-surge fuses, thermal cut-outs, fusible resistors and circuit breakers. All of these devices can be inserted in the primary or secondary side of the power supply, whether it be a switched-mode or linear supply. However, it is much better to *limit the current to a pre-determined maximum value*, as we did with linear regulators in Chapter 3, and similar circuits can be used with S.M.P.S.

The output current is controlled indirectly by sensing the voltage drop across a low-value series resistor. The resultant voltage is then used to bias-on a protection transistor, or circuit, which switches off or reduces the conduction of the regulator device. Figure 7.23 shows the application of this current limiting technique as used in a S.M.P.S.

The value of R_S and the voltage drop across it are kept small so as to minimize power dissipation. A typical value for V_S is 50 millivolts, thus keeping the maximum power dissipated in R_S down to about 1 per cent of the output power. However, this value varies with circuit design. The current limit amplifier overrides the output of the error amplifier and takes

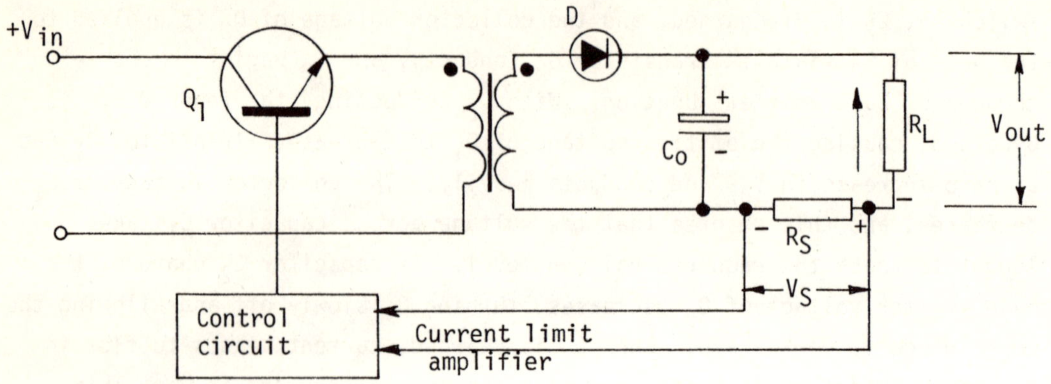

Figure 7.23 Current limiting

control of the pulse width. The duty cycle decreases as V_S increases, and V_{out} is reduced accordingly.

A typical current-limiting characteristic curve is shown in Figure 7.24, and with careful design the short-circuit current can be reduced to a safe level.

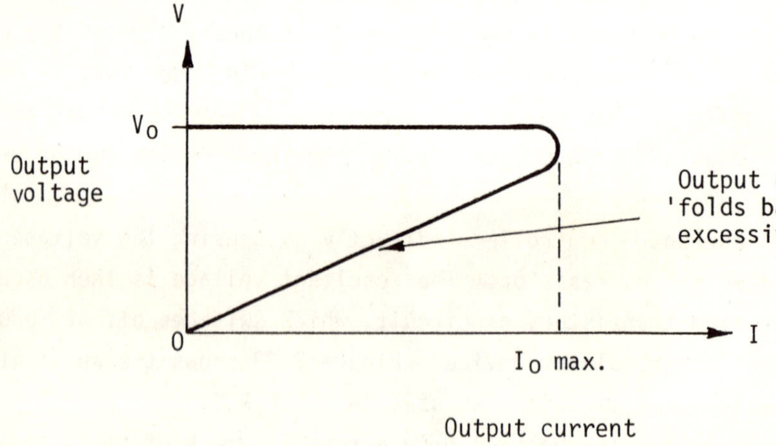

Figure 7.24 Output current foldback

Another way to protect against overcurrent, is to use the sensed voltage across R_S to latch up the oscillator, and so remove the drive from the switching device. The output of the power supply is then removed very rapidly.

7.6.3 INPUT SURGE CURRENT PROTECTION

Many switched-mode power supplies are operated directly off the 240 V A.C. line with capacitor input filters. Therefore, some method of preventing the rectifiers from being destroyed by a high 'surge current' is usually required. Figure 7.25 shows a circuit that is capable of limiting the surge current, and then allowing full voltage to the rectifiers.

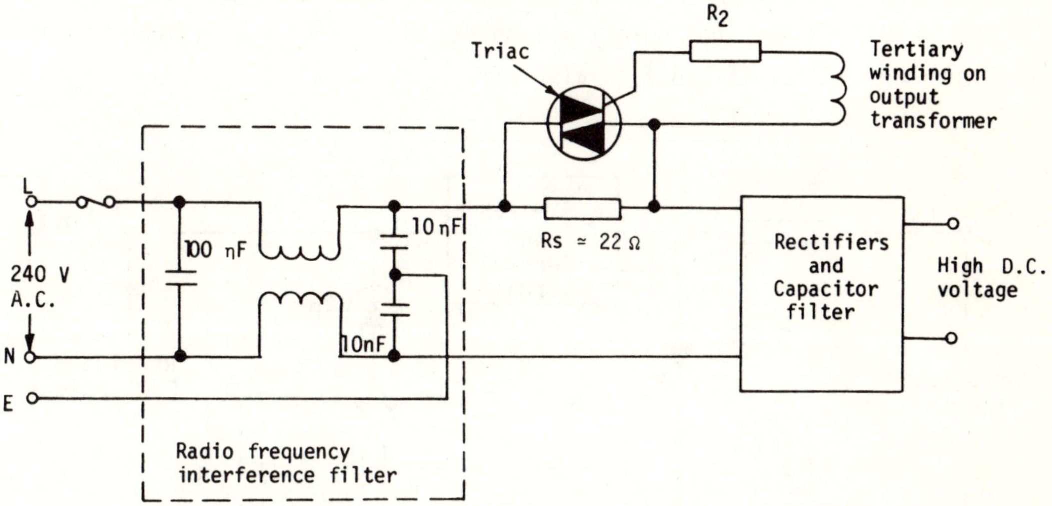

Figure 7.25 Surge current limiting (plus R.F.I. filter in the input line)

The series resistor (R_S) is used as the element for surge-current limiting, and hence protects the rectifier diodes at switch-on. Once the input filter capacitor has charged and the switched-mode circuit is operating, a voltage is induced in a tertiary winding on the output transformer. This voltage switches on a triac, which shunts R_S and virtually eliminates any power dissipation in R_S. The triac is not brought into conduction until the input capacitor is charged, and the surge current associated with initially charging the capacitor is therefore very low. Resistor R_2 limits the gate current of the triac.

7.6.4 OVER-VOLTAGE PROTECTION

Some form of over-voltage protection is necessary because an excessive output voltage may cause damage to integrated circuits and other semi-conductor devices in the load circuit. One very effective method of protecting against over-voltage is 'electronic crowbar protection'. This is basically a circuit that applies a short circuit to the power supply if the output voltage exceeds its rated level. (Figure 7.26 shows a 'crowbar' circuit.) The excessive current can (1) blow a fuse, (2) operate a cut out, or (3) initiate current foldback as discussed in Section 7.6.2, depending on the circuit connection. The avalanche voltage of the zener diode is chosen so that it does not normally conduct; therefore, the silicon controlled rectifier (S.C.R.) is also off.

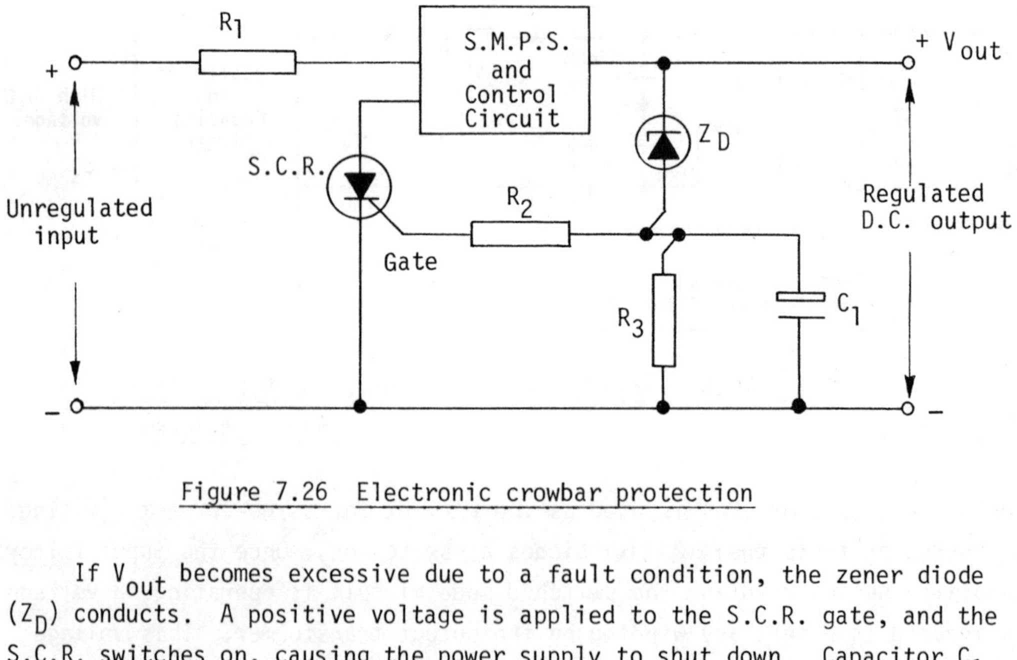

Figure 7.26 Electronic crowbar protection

If V_{out} becomes excessive due to a fault condition, the zener diode (Z_D) conducts. A positive voltage is applied to the S.C.R. gate, and the S.C.R. switches on, causing the power supply to shut down. Capacitor C_1 absorbs any sudden fluctuations (transients) in the load circuit that may inadvertently trigger the S.C.R.

With integrated controllers, a simple comparator can be used to keep a check on the output voltage condition. The output is monitored, and a portion of it is fed back to a comparator to be compared with a reference voltage. At a predetermined voltage 'difference', the comparator triggers the protection circuit, and the switching drive waveform is either reduced in duty cycle or taken to zero, which switches the circuit off. Figure 7.27 shows this circuit arrangement.

7.6.5 SWITCHING TRANSISTOR PROTECTION

In practice, the switching transistor cannot turn off the instant the drive is removed. As soon as the collector current falls, the collector voltage will start its swing to the opposite value. It depends on the circuit configuration as to whether it swings positive or negative. If there is a negative input voltage to the emitter, the collector voltage swings positive on switch-off. If there is a positive input voltage to the collector (as shown in the figures in this text), the emitter voltage swings negative on switch-off.

Whichever the case, precautions are required to prevent the transistor from dissipating too much power during the transition. This is termed keeping the transistor in its 'safe operating area' (S.O.A.R.). The components C_1, D_1, and R_1 in Figure 7.28 are included in circuits for this purpose (some variations will be encountered in different designs).

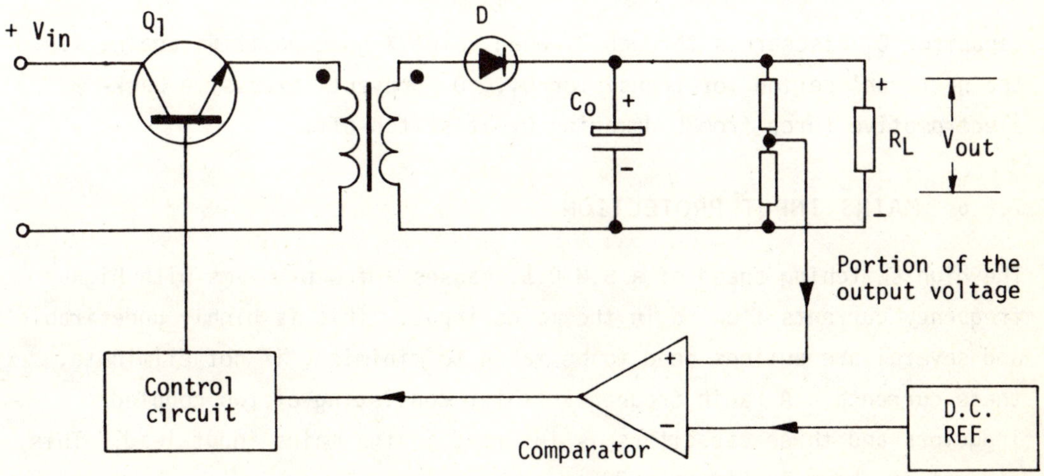

(The D.C. reference and comparator could be included in the I.C.)

Figure 7.27 Voltage protection using a comparator

The small value of inductance (L) is inserted between the transistor emitter and the transformer primary to prevent reflected spikes from the transformer secondary circuit reaching Q_1 at switch-on. When the switching transistor Q_1 turns off, Point X swings negative, which causes D_1 to conduct. The transformer primary (T_p) and C_1 form a series-tuned circuit with *low* impedance, and the current surges into C_1, thereby reducing the current through Q_1. This prevents the collector-emitter voltage of Q_1 from rising too quickly at switch-off, hence reducing the power dissipated by Q_1.

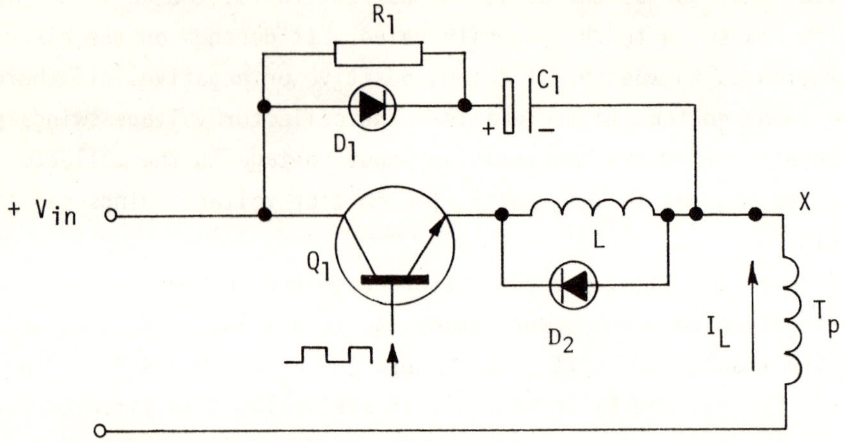

Figure 7.28 Transistor protection

Capacitor C_1 discharges through R_1 when Point X goes positive again, i.e. the next 'on' period for transistor Q_1. D_2 prevents excessive back-electromotive force from L damaging Q_1 at switch-off.

7.6.6 MAINS INPUT PROTECTION

The high switching speed of a S.M.P.S. causes a few problems with high-frequency currents flowing in the mains input. This is highly undesirable and several precautions need to be taken to minimize, if not eliminate, these currents. A radio frequency filter consisting of two coupled inductors and three capacitors is inserted in the mains input lead. This circuit is shown in Figure 7.29.

Capacitors C_1, C_2 and C_3 form the filter circuit with inductors L_1 and L_2. The inductors are mounted on ferrite cores; they are connected so that the two leakage fluxes are in opposition in order to prevent magnetic radiation. A further precaution is to by-pass the rectifier diodes with small value capacitors (C_4 - C_7 in Figure 7.29). These capacitors suppress

H.F. and V.H.F. radiation caused by the diodes switching on and off at the mains rate.

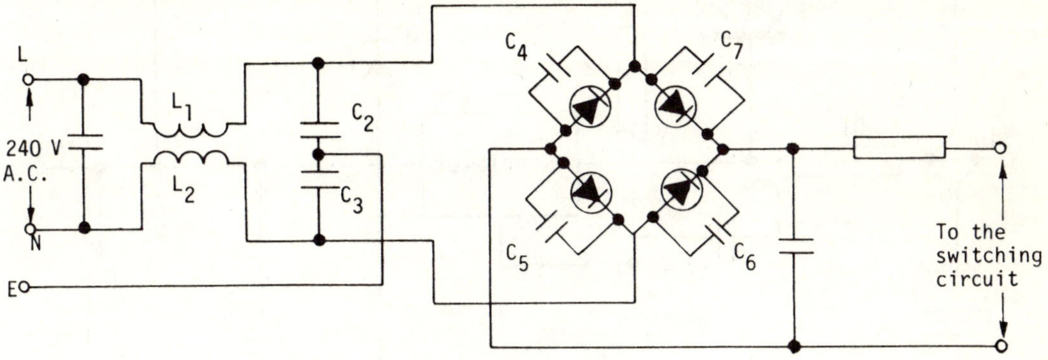

Figure 7.29 Mains input protection

7.7 THREE-TERMINAL REGULATORS IN SWITCHED-MODE CIRCUITS

Low cost adjustable switching regulators can be made using three-terminal I.C. regulators as the control device. The simplest circuit configuration is shown in Figure 7.30.

V_{in} is a rectified low-voltage input. Voltage divider R_2 and R_3 is the feedback loop from the output. It sets the voltage on the adjustment pin of the I.C., which determines the output voltage (refer to Section 4.8).

Transistor Q_1 is the switching element, inductor L is the energy storage device, and diode D provides the current path for the energy returning from L when Q_1 is off. C_1 is the normal input capacitor, C_2 is the output storage capacitor and is much larger than normal, and C_3 is included to minimize the ripple level at the junction of R_2 and R_3 - i.e. filters any variations in the feedback voltage.

When the power supply is switched on, the load draws current through the regulator. This turns Q_1 on and allows current to pass through the inductor, L. As the current through L increases, the regulator supplies less and less current to the load and finally turns Q_1 off. The field around L collapses, the polarity of the voltage across L reverses, forward biasing the diode D and current continues in the output, even though Q_1 is off. Once the energy in L is spent, the load draws current through the regulator and turns transistor Q_1 on again. The action then repeats itself.

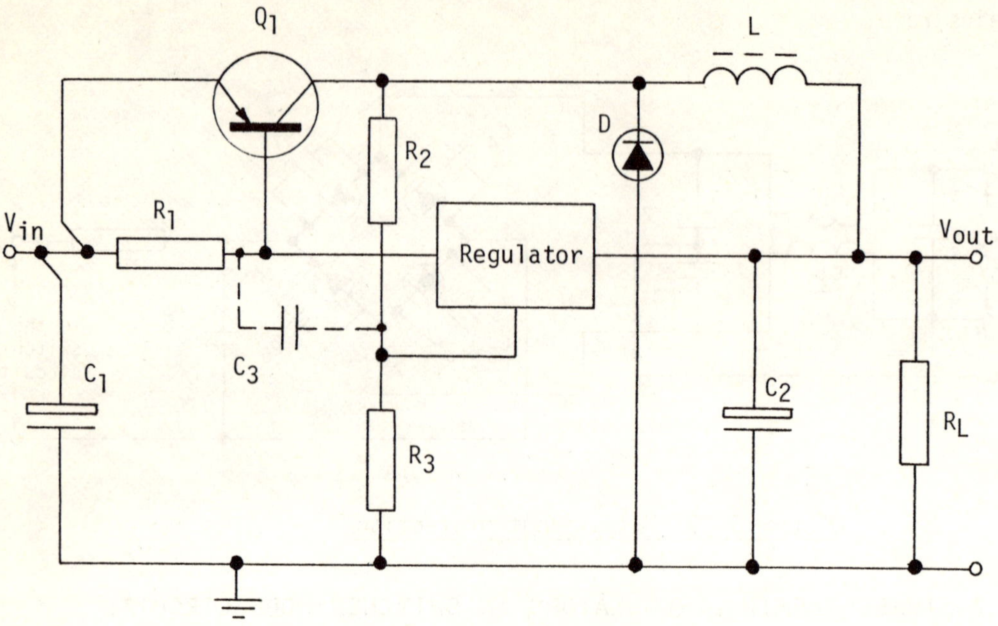

V_{in}	= 10 V	R_1	= 500 Ω
V_{out}	= 5 V	R_2	= 1 Ω
I_{out}	= 300 mA	R_3	= 10 Ω
L	= 500 μH	C_1	= 2.2 μF
f	= 37 kHz	C_2	= 100 μF
M:S Ratio	= 1:1	C_3	= 100 μF
Q_1	= 2N2905	D	= 1N5807

Figure 7.30 Switching regulators using a three-terminal regulator

A further development is to make the output adjustable by varying the voltage on the adjustment pin. A circuit using this technique is shown in Figure 7.31. It operates in a similar manner to the circuit in Figure 7.30, but uses an adjustable voltage power regulator (LM317) which enables a much higher output current to be obtained.

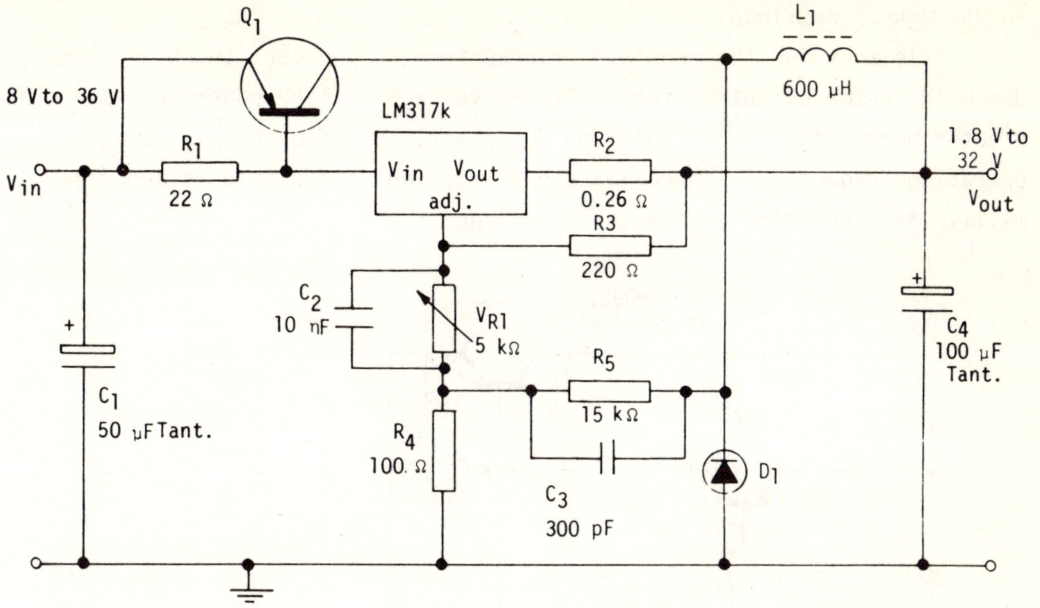

Q_1 = 2N3792 or equivalent

D_1 = 1N3880 or equivalent

L = 600 μH (60 turns)

Figure 7.31 Low-cost 3 ampere switching regulator for experimenters

A power P.N.P. is used as the switch, driving the L_1 C_4 filter. Positive feedback is applied to the LM317 via R_5. When transistor Q_1 switches, a small square-wave voltage is generated across resistor R_4. This is level-shifted and applied to the adjustment terminal of the regulator by VR_1 and C_2, causing it to switch ON or OFF. Negative feedback it taken from the output through resistor R_3. Capacitor C_3 acts as a speed-up capacitor, increasing switching speed, while R_2 limits the peak drive current to Q_1. A blow-out-proof switching regulator is shown in Figure 7.32.

The PNP switching transistor has been replaced by a PNP-NPN combination using LM395s as the NPN transistors. The LM395 is an I.C. which acts as a NPN transistor with overload protection. Included in the I.C. is current-limiting, safe-area protection, and thermal-overload protection (as discussed in Chapter 4), making the device virtually immune to any type of overload.

Efficiency for the regulator ranges from 65 per cent to 85 per cent, depending on the output voltage. At low voltages, fixed power losses are a greater percentage of the total output power, so efficiency is lowest. Operating frequency is approximately 30 kHz, and the ripple is about 150 millivolts, depending on the input voltage.

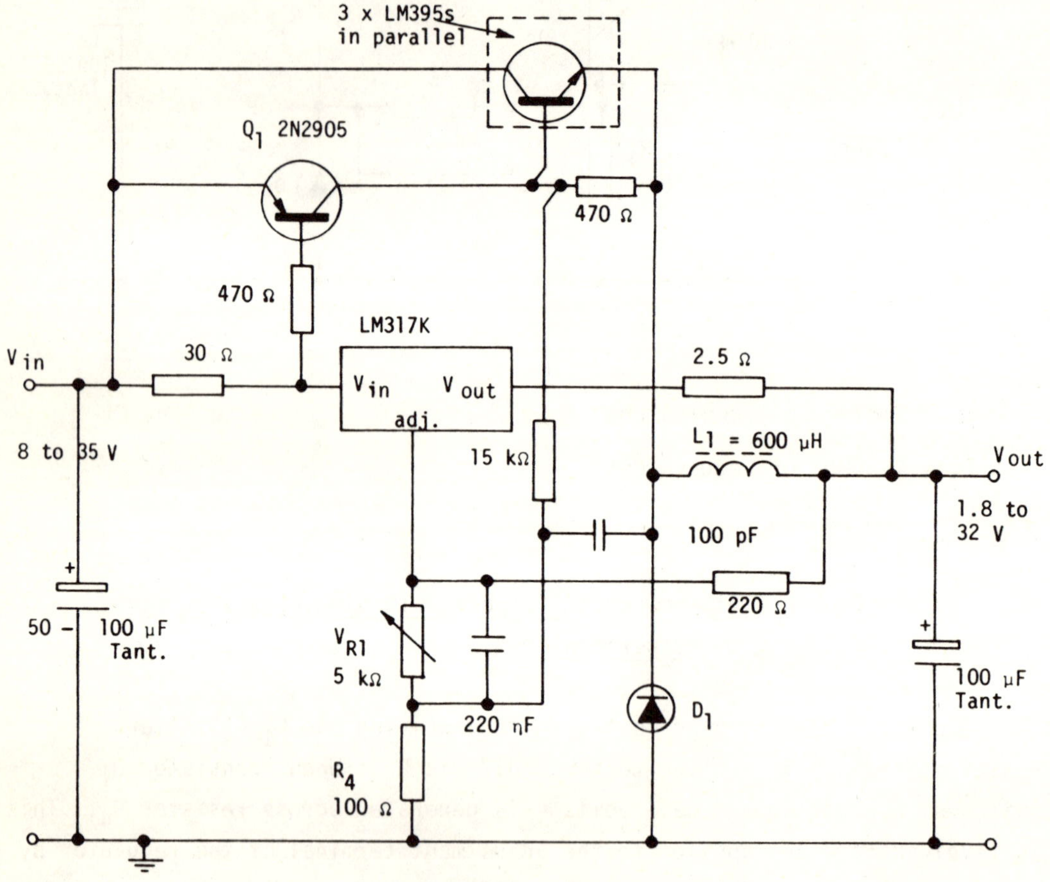

Figure 7.32 4 ampere switching regulator with overload protection

One of the more unique applications for these switching regulators is as a tracking pre-regulator. The only D.C. connection to ground on either circuit is through the 100 Ω resistor, R_4. Instead of tying this resistor to ground, it can be connected to the output of a linear regulator, so that the switching regulator maintains a constant input-to-output differential on the linear regulator. This technique was discussed in Section 7.2.1. The switching regulator would be set to hold the input voltage to the linear regulator about 3 volts higher than the output. The operating conditions of the linear regulator are now held constant, because any fluctuations in input or output voltage are taken care of by the switching pre-regulator.

7.8 SELF-EVALUATION QUESTIONS (ANSWERS IN APPENDIX A)

1. When making a comparison between linear and switching regulators, what are the three main requirements for an 'ideal' system?

2. What are the main advantages of a switched-mode power supply?

3. In your opinion what are the disadvantages of a switched-mode supply? How can they be overcome?

4. What danger is encountered with some S.M.P.S. systems that is not normally encountered with power supplies?

5. Is there any special requirement in output capacitors with an S.M.P.S.; if so, what is it?

6. Draw the circuit arrangement for a basic series-switching regulator and explain its operation. Show drive voltage waveforms.

7. What is the main difference in circuit configuration between a step-up and a step-down S.M.P.S.?

8. Which of the two circuits in Question 7 exhibit the least noise characteristics? Why is this so?

9. (a) Can a S.M.P.S. be used to provide mains isolation?
   (b) Which circuit would you use?
   (c) Draw the circuit and explain its operation.

10. In an 'inverting' S.M.P.S., where is the energy stored when the switching element is turned on? What is the purpose of the diode?

11. What is the difference between a 'forward' and 'flyback' S.M.P.S.? Which one has the highest power capability?

12. Draw a block diagram of the control circuitry required for an S.M.P.S., and briefly explain the function of each block.

13. Why is a 'soft-start' circuit required in an S.M.P.S.?

14. Briefly explain how a soft-start circuit works.

15. Explain two methods that may be used to 'current-limit' an S.M.P.S.

16. What is meant by 'crow-bar' protection? How does the circuit work?

17. Why do precautions have to be taken to protect the switching transistor when it turns off?

18. Draw a typical circuit you would expect to find in the A.C. input of an S.M.P.S. What is its function?

8 Fault Finding and Test Circuits

8.1 INTRODUCTION

A high percentage of faults in electronic equipment are usually tracked down to the power supply. After all, the power supply is probably the hardest working area of any piece of equipment. It has to supply sufficient voltage and current to keep the device it is connected to working satisfactorily. It has to suffer the stress of abuse, because it must occasionally supply excess current due to a fault condition in its load, or dissipate excess power due to a fault condition in its input circuit. Under fault conditions, the life of a power supply is not a happy one, and if insufficient precautions are taken to protect it, it will fail - through no fault of its own.

The basic assumption to make when servicing a power supply unit is that a fault has occurred in one of its components, and when the faulty component is located and replaced, the power supply will function correctly again. However, one thing that must be considered is: did the power supply fail by itself, or was the failure induced by a fault in the circuits connected to it? This chapter outlines some fault-finding techniques and procedures for locating and rectifying failures in the circuits discussed in the preceeding chapters.

8.2 CIRCUIT MEASUREMENTS AND COMPONENT FAULTS

8.2.1 VOLTAGE MEASUREMENTS

These measurements are made to check the A.C. input and/or D.C. output voltages of the power supply. They are the simplest to perform and do not necessitate any component removal. The instrument used is a voltmeter with a high sensitivity (greater than 20,000 ohms per volt), so that the circuit under test is not loaded excessively by the meter. Remember to choose the range on the meter to suit the voltage you expect to find in the circuit. If in any doubt, start on the highest range to avoid damaging the meter movement.

8.2.2 RESISTANCE MEASUREMENTS

These are not usually made with the components in circuit, since parallel paths can often give a misleading reading. However, when servicing power supplies, or circuits connected to power-supply outputs, checking the output line for any obvious short circuit can be useful for diagnosis. *Remember:* all resistance measurements in a circuit must be carried out with the equipment unplugged and with any large capacitors discharged to avoid damaging the meter. A further point to watch is the polarities on the test probes when measuring areas containing diodes and transistors. The potential at the test probes may easily forward bias the rectifier diodes in a power supply and indicate a false result.

8.2.3 CURRENT MEASUREMENTS

These are usually only made when unusual faults occur, because it is often difficult to break the circuit and insert the meter - especially on printed circuit boards. However, they are an essential measurement in the final testing of the power supply in order to determine the D.C. output current available.

8.2.4 OSCILLOSCOPE MEASUREMENTS

An oscilloscope can be an invaluable piece of test equipment and can quickly give a visual indication of a circuit failure. For example, if the waveform of a half-wave circuit appears at the output of a full-wave rectifier circuit, then the area of the fault is located instantly. It is essential to use an oscilloscope in the final testing of the power supply to check the amplitude of the A.C. ripple on the D.C. output voltage.

8.2.5 COMPONENT FAULTS

Capacitors usually have one of three faults, as discussed in Chapter 2:

(a) Open circuit - it has little if any capacitance, and no meter deflection is given when checked with an ohmmeter.

(b) Short circuit - in this case it will read zero ohms when checked with an ohmmeter.

(c) Leaky - in this case it will have a resistance value that is considerably less than infinity.

Resistors can change in value, due to excessive current being passed through them, or they may be completely open-circuit.

Transistors suffer from the same problem as *diodes*, and these were discussed in Chapter 1. The usual problems encountered with semi-conductor devices are:

(a) Short-circuited junction, caused by a high-voltage surge.

(b) Open-circuited junction, usually caused by a heavy overload, and hence excessive current.

(c) High-leakage current, usually accompanied by low-gain or a high-noise level.

Transformers and *inductors* generally develop an open circuit, but sometimes a shorted turn may be found. The testing procedure was discussed in Chapters 1 and 2. A device that may prove useful when checking coils is a shorted turns tester. This is an oscillator circuit and the coil under suspicion is put across the winding. If there is no fault, the frequency and current drawn will only vary slightly. However, if it has a shorted turn, the oscillator will stop, and a large current will be drawn through the circuit.

8.3 SAFETY PRECAUTIONS

There are two aspects to consider when it comes to safety in the workshop or on the job:

(1) Safety of individuals - this means your safety, and in most cases, you are responsible for it.

(2) Safety of the boss's equipment - he doesn't want his precious, and probably expensive, test equipment damaged.

Electricity is a silent medium and should be treated with caution and respect. If you are careless when handling electronic equipment, quite often the first thing you know about it is a terrible sensation up your arm or across your chest. *Very small electric currents of 100 milliamps, often less, through the body are potentially lethal.* It is important to remember that it is not a matter of how much voltage is required to produce this current; rather, the crucial factor is your body resistance. From the famous law, I = V/R, we can determine this important information. If your body resistance is high, say 10,000 ohms, then you would require 1000 volts to produce 100 mA of current. As your body resistance decreases, so does the amount of voltage required to kill you.

Technicians and servicemen who work on 'low-voltage' equipment are often lulled into a false sense of security. In a situation where body resistance might be as low as 100 ohms, then 10 volts is potentially lethal. Such a situation might be unlikely in practice, but consider this: if you suddenly switch from servicing a low-voltage power supply to a high-voltage supply, will you also remember to switch your safety precautions? The probable answer is 'No'. Therefore, you are in danger of becoming a statistic.

In electronics you are likely to encounter power supplies of all shapes and sizes. You may even experience a 'transformerless' circuit or a S.M.P.S.

working directly from the A.C. (as in Figure 7.2). With these circuits there is always the possibility of the 'live' and 'neutral' wires of the A.C. input becoming swapped over. If this happens in a 'transformerless' circuit, then the *chassis of the equipment is 'live'*. Therefore, adopt a sensible attitude to safety, do not be careless, and do not fall into bad habits, such as unsoldering components from energized circuits.

Always use an isolation transformer between the A.C. mains and the equipment you are working on. This will prevent a direct short circuit being placed across the mains and will protect both the technician and the test equipment. If you are building a workbench, an isolation transformer is an essential piece of equipment for the workshop. It must be wired between the A.C. mains input and the power outlets on your workbench. The volt/amp rating of the isolation transformer must be sufficient to handle the power requirements of the equipment you intend to service on the workbench and also the test equipment you intend to use. Figure 8.1 shows a block diagram of the workbench arrangement.

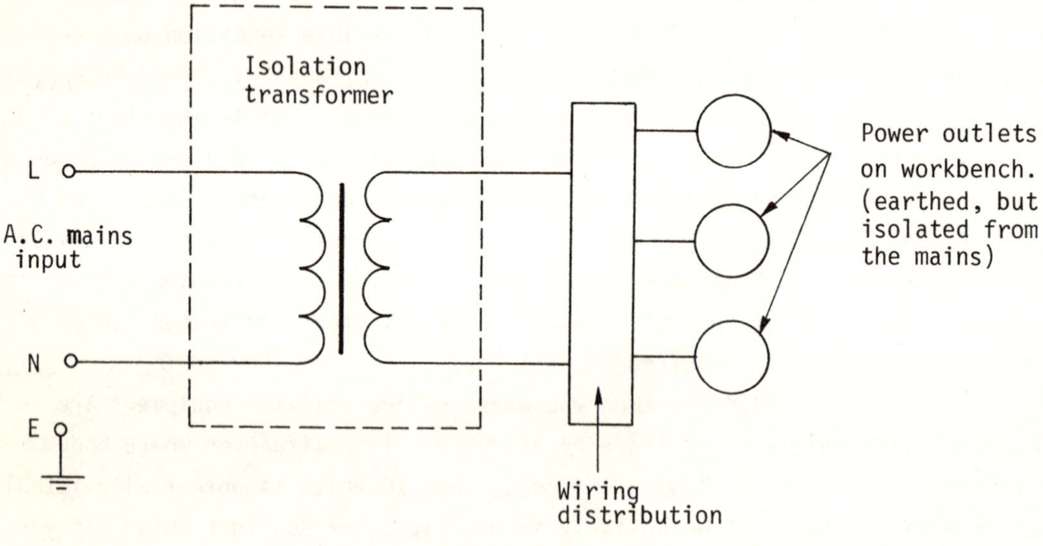

Figure 8.1 Isolating the workbench from the A.C. mains input

Equipment damage usually comes under the heading of carelessness or sloppy practices. Before using a meter, *check the position of the range selector*. Be sure you do not check the continuity of a coil and then measure a D.C. voltage. Keep the workplace tidy. There is no excuse for a meter being knocked off the bench because it got caught up in a power lead when a piece of equipment was being removed after servicing.

8.4 COMPONENT REPLACEMENT AND CIRCUIT-BOARD REPAIRS

8.4.1 COMPONENT REPLACEMENT

When replacing faulty components in power-supply circuits it is important to obey two rules:

(1) Replace the faulty component with an identical or equivalent component.

(2) Position the replacement component in the exact physical location from which the faulty component was removed.

It is important to make sure that the replacement component is the same or equivalent because some power supplies use special components. A three-terminal regulator circuit uses tantalum capacitors. They have a smaller physical size, but they also have a much higher frequency capability than standard electrolytics. The capacitors used in an S.M.P.S. must have a low equivalent series resistance, (E.S.R.). Likewise, transistors and diodes used in switching circuits must have fast rise and fall times. Some power supplies will use wire-wound resistors in place of filter chokes - they exhibit a small amount of inductance and aid filtering. The D.C. working voltage of replacement components should also be watched. The working voltage of the replacement may be higher than the original, but never lower.

Concern with the exact positioning of the component may seem a little unnecessary. Some might say: if it is wired in, then it must work. True, but certain components such as filter chokes have quite a large magnetic field associated with their operation. When a power supply is designed, magnetic radiation is taken into account, and component are placed in the optimum working positions. If this position is shifted, interference may result, giving rise to hum or higher ripple amplitude. Fortunately, printed circuit boards have almost superseded the choice of locality: in these circuits the new part must go back where the faulty part came from. However, there are still a few custom-built power supplies around which do not use printed boards, and positioning of components can be an important consideration.

8.4.2 PRINTED BOARD REPAIR

When printed circuit boards are used in power supplies, excessive current can be a problem. The high current heats the printed copper track, and this can melt the adhesive with the result that the copper track lifts off the board. Small-scale repairs, such as resticking loose copper track or by-passing hair line cracks, can be economically carried out, but more complicated repairs often justify replacing the whole board. If the board has been subjected to a heavy current due to excessive load conditions, the latter is the more practical solution. New components should also be fitted to the board rather than removing them from the old board and resoldering them to the new board.

Loose copper tracks can usually be restuck to the board by carefully applying a quick-setting epoxy and holding the track to the board until the adhesive has set. Hair-line cracks can be by-passed with a short length of wire soldered to the copper track. The track must be clean, and the heat should be applied only as long as necessary. If a long 'jumper' wire is required, it should be insulated to prevent accidental short circuits to other parts of the circuit. Care should be taken when soldering, not to short out copper tracks that are in close proximity to the track being repaired.

8.4.3 COMPONENT REPLACEMENT ON PRINTED CIRCUIT BOARDS

As mentioned in the previous section, heat can melt the adhesive on the board and cause the copper track to lift off. Sometimes a component can be very stubborn during removal, and the heat applied to free it can become excessive. In this case, note the value and working voltage (if applicable) of the component, ensure that you have a replacement, and then cut the component from its mounting leads. The mounting leads are then left protruding through the top of the board. These can be cleaned, the new component leads trimmed to size and soldered to the protruding leads, using a minimum amount of heat.

Ideally the solder should be removed from the printed circuit and the new component fitted directly to the board. Special de-soldering irons are available which suck the solder away from the board, but these are relatively expensive. A mechanical spring-loaded 'solder-sucker' and 'solder-wick' (flux-coated copper braid) are also very good for removing solder from the printed circuit. However, whichever method is used, heat must be kept to a minimum to ensure that there is no lifting of the copper track.

8.5 FAULT-FINDING PROCEDURE AND TESTING TECHNIQUE

When a power-supply unit requires service, the fault must be isolated to a particular area (as with all electronic equipment), and a logical servicing procedure is required for efficient and economical servicing. General troubleshooting techniques are as follows, and these may be applied to power supplies.

1. *Basic assumption.* As with all radio, television and electronic servicing, the basic assumption is that only one fault is causing the malfunction of the power supply. When the fault is located, and the defect rectified, normal operation will be restored. If normal operation is not restored, a second source of trouble should be investigated.

2. *Isolate the defective section.* Refer to any service data available, and familiarize yourself with the circuit layout and characteristics. A visual inspection of the suspect section will reveal any obvious faults, such as broken printed track, broken components, burnt out resistors, and, if a valve unit, unlit valves. If the fault is not found visually, checks should be made at certain test points to determine the circuit D.C. voltage conditions.

3. *Isolate the defective stage.* When the defective section has been located, test equipment is used to isolate the exact fault locality.

4. *Isolate the defective component.* After the faulty stage has been located, voltage and resistance checks are used to locate the defective component. Substitution of known good components for suspect components may be required to help in the diagnosis, especially on more complex circuits.

5. *Correct the defect.* The defective part is replaced with the exact same or equivalent component.

6. *Preventative maintenance.* When operation has been restored to normal, other defects that may cause trouble later, such as cracked resistors, overheated and leaky capacitors, should be replaced. If it is a valve unit, valves that flash or arc when lightly tapped are also a potential source of trouble.

When defective components are replaced, bear the following points in mind:

(1) Keep the same positions for earth return and connecting leads.

(2) Lead-dress is critical with high frequencies to prevent feedback, also at high voltages to prevent arcing.

(3) Resistors may be replaced with ones of a higher wattage, but not lower.

(4) Capacitors may be replaced with the same type of higher working voltage, but not lower.

(5) Mica, ceramic, or polyester capacitors should not be replaced with paper capacitors, as their inductance will alter the coupling or bypassing at high frequencies.

When the circuit operation appears to be satisfactory, the power supply parameters (operational statistics) should be tested.

8.5.1 POWER SUPPLY TESTING

The main points which should be tested after a power supply has been repaired are:

(1) D.C. output voltage.

(2) D.C. output current available.

(3) Ripple voltage amplitude at the output.

(4) Voltage regulation, or current regulation if a constant current supply.

(5) Circuit stability with variations in the mains input voltage.

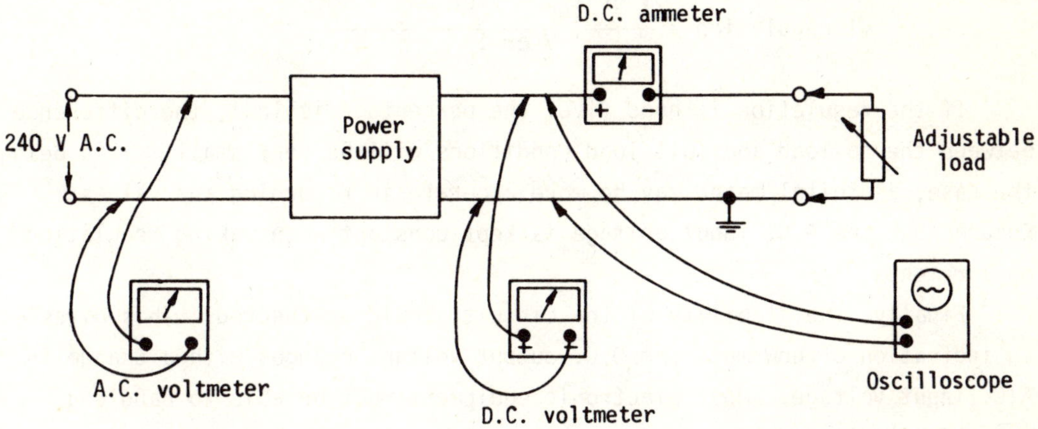

Figure 8.2 Power supply under test

A power supply can be tested for satisfactory operation by the connection of an A.C. voltmeter, a D.C. voltmeter, a D.C. ammeter and an oscilloscope as shown in Figure 8.2. The A.C. input voltage should be checked and noted. Unfortunately, the only means available for adjusting the A.C. input is to connect the unit to the mains supply via a 'variac' (adjustable auto-transformer). The input voltage should be adjusted to the nominal voltage setting on the power supply: 220 V, 230 V or 240 V.

The D.C. voltage should be measured under full load conditions. With the load set to maximum resistance, note the output voltage. Then slowly decrease the resistance until the current in the load reaches the maximum value; there should be little change in the output voltage. Take care not to short circuit the output voltage by adjusting the variable resistance to zero ohms. The power supply should now be delivering maximum current at the rated output voltage.

The peak-to-peak ripple amplitude can be checked with the oscilloscope. Set the oscilliscope to A.C. and adjust the vertical amplitude to a sensitive range. The peak-to-peak amplitude depends on the type of supply and filtering arrangement, but should be quite low - typically in the range of 5 mV to 20 mV.

Voltage regulation is determined by measuring the output voltage no-load and full-load conditions. Likewise, with current regulation, the output current is measured under no-load and full-load conditions as explained in Chapters 2 and 3. The formula used expresses the regulation as a percentage:

$$\% \text{ regulation} = \frac{V \text{ or } I_{NL} - V \text{ or } I_{FL} \times 100}{V \text{ or } I_{FL}}$$

If the regulation is good (i.e. the percentage is low), the difference between the no-load and full-load conditions will be very small. This being the case, a digital meter may be more accurate in recording the values. Ensure that the A.C. input voltage is kept constant when making regulation measurements.

Finally, the stability of the circuit should be checked. This gives an indication of how much the D.C. output voltage changes with a change in A.C. input voltage. Most electronic equipment must be able to handle a swing in A.C. input voltage of approximately $\pm$ 8 per cent. This means the A.C. input can vary from 220 volts to 260 volts, about an average of 240 V A.C. The input voltage is adjusted by varying the variac and monitoring the A.C. voltmeter across the input. As the input voltage is varied from 220 to 260 volts, the D.C. output voltage is noted, with the load set at maximum. If a swing of $\pm$ 8 percent at the A.C. input causes the D.C. output to vary by 50 mV from a rated value of 12 V, then the output changes by:

$$\frac{50 \times 10^{-3}}{12} \times 100 \simeq 0.4$$

The stabilization is expressed as a ratio: $\pm$ 8 per cent is a total possible variation of 16 per cent. Therefore, a variation of 16 per cent at the input causes a variation of 0.4 per cent at the output. The line stabilization is thus 16 : 0.4, or 40 : 1.

8.5.2 FAULT CONDITION AND SYMPTOM TABLES

In Figure 8.3, a fault may be in the transformer, the rectifier diodes, the filter section, or the regulator. Diagnosis should start with a few obvious checks. First, ensure that the A.C. mains is plugged in and switched on. Check the D.C. output voltage: if no output, check that the A.C. is reaching the transformer secondary. If no A.C. at the secondary, three faults are possible:

(1) Open-circuit primary or secondary.

(2) Open-circuit mains wire inside the plug.

(3) Blown fuse.

To check these, unplug the mains and perform continuity checks with an ohmmeter. Do not rely on a visual check for the fuse; use the ohmmeter.

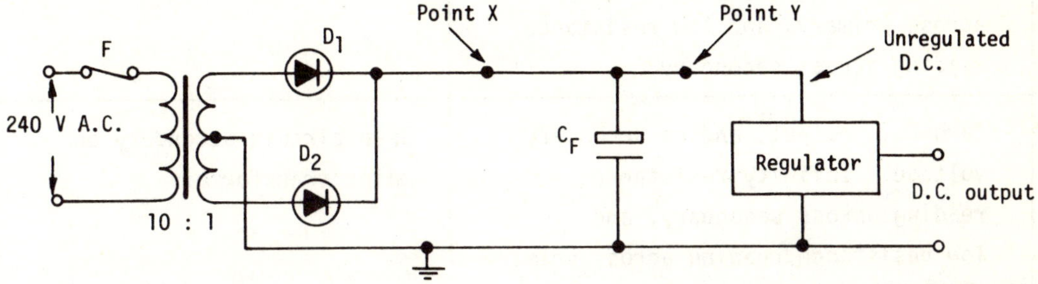

Figure 8.3 Power supply circuit

If the fuse has blown, a fault condition in the power-supply circuit has caused it. This fault must be located and repaired prior to fitting a new fuse.

Ensure capacitor C_F is discharged and perform a continuity check from the junction of D_1 and D_2 to earth (watch the polarity of the test probes!). If a short circuit or very low resistance results, disconnect the output line just before the filter capacitor at Point X and test again. This isolates the fault area; if the same reading results, the fault is either in the rectifier diodes or in the transformer; if there is a different reading, the fault is either the capacitor C_F or the regulator. Disconnect the circuit at Point Y and perform continuity checks across C_F and the regulator. Should the capacitor check out satisfactorily, the problem will be in the regulator circuitry. Voltage readings are then taken around the various test points in the regulator to determine the faulty component.

Typical faults and their symptoms in the area from the A.C. mains input up to the regulator input are listed in the table in Figure 8.4.

The next area of concern in the power-supply unit is the regulator itself. A series regulator circuit is shown in Figure 8.5. Fault finding in the regulator is usually straightforward. Typical working D.C. voltages

SYMPTOMS	FAULT
<u>Transformer faults</u> Zero D.C. output, and no secondary voltage. Low resistance readings on both primary and secondary.	Open-circuit A.C. input lead or blown fuse.
Zero D.C. output, and no secondary voltage. Infinity resistance reading across primary, and low resistance reading across secondary.	Open circuit primary on mains transformer.
Zero D.C. output, and no secondary voltage. Infinity resistance reading across secondary, and low resistance reading across primary	Open circuit secondary on mains transformer.
Low D.C. output, and transformer excessively hot. (Fuse may also blow)	Shorted turns on mains transformer primary or secondary.
<u>Rectifier faults, (full-wave and bridge)</u> Low D.C. output with 50 Hz ripple. Poor regulation.	Open circuit rectifier diode.
Fuse blown due to excessive current. Resistance check on transformer O.K.	Short circuit rectifier diode.
Circuit operates O.K., but D.C. output is lower than expected.	High resistance diode. (High forward voltage drop)
<u>Filter capacitor faults</u> Low D.C. output, with high level of ripple. Very poor regulation.	Filter capacitor open circuit.

Figure 8.4 Table of faults and symptoms in the transformer, rectifier and filter stages of a power supply

Figure 8.4 continued

Fuse blown. Resistance of the unregulated D.C. line is low in both directions.	Filter capacitor short circuit.
Lower D.C. output, with increase in ripple level. Poor regulation. (Equipment function may be unstable)	Filter capacitor leaky.

are compared to the actual circuit voltage, and a diagnosis made. The voltage reference points are referred to as 'test points' (TP).

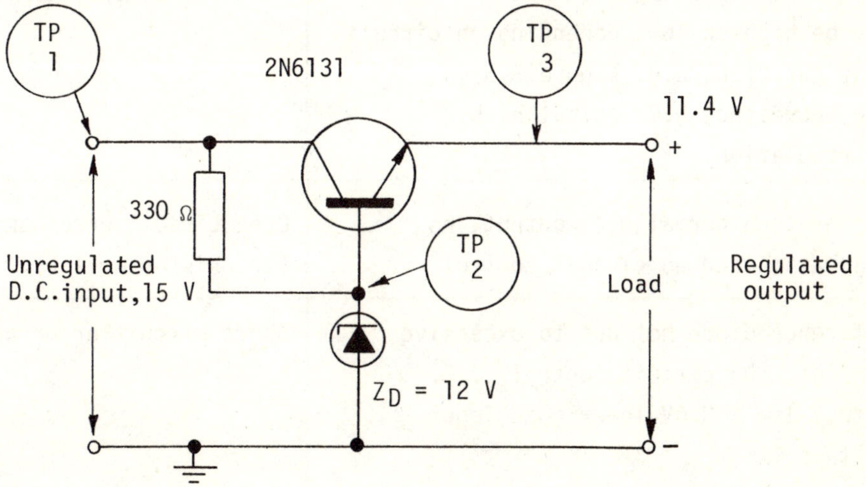

Figure 8.5 Basic series regulator with test points

The test-point voltages are:

TP 1 = 15 volts

TP 2 = 12 volts

TP 3 = 11.4 volts

A table of faults in regulator circuits and their symptoms is shown in Figure 8.6.

SYMPTOMS	FAULTS
Zero D.C. output, and the unregulated D.C. input slightly higher than normal. (no current drawn)	Series transistor base-emitter open circuit.
Input and output voltages equal, may blow a fuse, or destroy the transistor due to excessive current.	Series transistor collector-emitter short circuit.
Low D.C. output, series transistor hot because $V_{CE} = V_{in} - V_{out}$	Zener diode short circuit.
No reference voltage. Output voltage may be high or low, depending on circuit configuration, (e.g. Figure 8.5), V_{BE} increased, D.C. output high. No regulation.	Zener diode open circuit.
Higher than normal D.C. output, no regulation and no circuit control.	Open circuit error amplifier (transistor or I.C.)
Reference diode hot due to excessive current. No circuit control, D.C. output low : 0.6V lower than Zener voltage.	Short circuit error amplifier.

Figure 8.6 Table of faults and symptoms in regulator circuits

8.6. FAULT DIAGNOSIS

In this section we will discuss the diagnosis of faults from information that is known about the D.C. circuit conditions at the test points.

8.6.1 FAULT DIAGNOSIS ON A SERIES REGULATOR

The first circuit is a basic series regulator with an error amplifier. Its operation was described in Section 3.4, and the circuit under investigation is shown in Figure 8.7. The normal D.C. voltages are measured with a 30,000 Ω/volt multimeter.

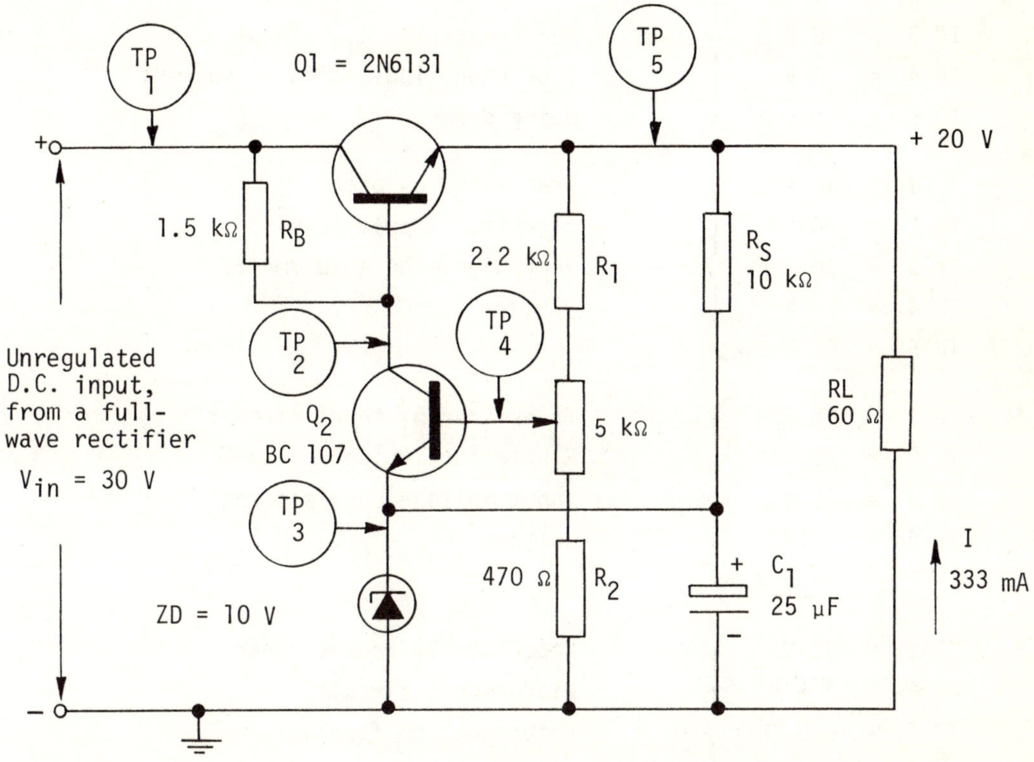

Figure 8.7 Adjustable series regulator with test points

Normal working conditions

TP 1 = 30 volts

TP 2 = 20 volts

TP 3 = 10 volts

TP 4 = 10.8 volts

TP 5 = Adjustable voltage

(Ripple level = 80 mV p-p. TP 5 set to 20 V for this exercise.)

Test point voltages under fault conditions

1. All test point voltages are 0 volts. Therefore the fault must exist in a stage prior to the regulator, i.e. transformer, rectifier, filter, or fuse.

2.

TP 1 = 33 V		R_B open circuit. No initial forward bias for transistor Q_1. Input D.C. high, indicating no current being drawn.
TP 2 = 0 V		
TP 3 = 0 V		
TP 4 = 0 V		
TP 5 = 0 V		

3.

TP 1 = 30 V		Transistor Q_2 open circuit. Ripple level very high. No adjustment.
TP 2 = 24 V		
TP 3 = 10 V		
TP 4 = 13 V		
TP 5 = 23 V		

4.

TP 1 = 28.5 V		Transistor Q_1 short circuit. Ripple level high. Input voltage equals output voltage.
TP 2 = 21 V		
TP 3 = 10 V		
TP 4 = 11.2 V		
TP 5 = 28.5 V		

5.

TP 1 = 32 V		Capacitor C_1 or the zener diode short circuit. No reference, no feedback voltage, output very low.
TP 2 = 2.5 V		
TP 3 = 0 V		
TP 4 = 0 V		
TP 5 = 2.0 V		

6.

TP 1 = 30 V		Zener diode open circuit. High ripple voltage. Reference too high, Q_1 turned on harder.
TP 2 = 24 V		
TP 3 = 22 V		
TP 4 = 13 V		
TP 5 = 23 V		

7. TP 1 = 33 V
TP 2 = 33 V
TP 3 = 0 V
TP 4 = 0 V
TP 5 = 0 V

Transistor Q_1 base-emitter open circuit. Input voltage increased. TP 1 and 2 equal, no I through R_B to Q_1 base.

8. TP 1 = 29.5 V
TP 2 = 24 V
TP 3 = 10 V
TP 4 = 1 V
TP 5 = 23 V

Resistor R_1 open circuit. High ripple. Input D.C. low, indicating higher current being drawn. No drive to Q_2 base.

9. TP 1 = 30.5 V
TP 2 = 11.5 V
TP 3 = 10 V
TP 4 = 11 V
TP 5 = 11 V

Resistor R_2 open circuit. Ripple level reduced. Q_2 on harder, and Q_1 turned off.

8.6.2 FAULT DIAGNOSIS ON A CURRENT-LIMITED SERIES REGULATOR

The next circuit to consider is a current-limited regulator. Figure 8.8 shows the circuit configuration that was discussed in Chapter 3. The output voltage is 20 volts at a maximum useable current of 500 mA.

Normal working conditions

TP 1 = 30 volts
TP 2 = 20 volts
TP 3 = 10 volts
TP 4 = 10.8 volts
TP 5 = 20 volts
TP 6 = 20 volts
Ripple voltage level, 20 mV p-p
Maximum current, 500 mA

Test point voltages under fault conditions

1. The value of R_L is decreased, and therefore excess current is demanded. (I = 570 mA)

TP 1	=	30 V	The excessive current flow through R_{CL} increases the voltage across Q_3 base-emitter junction. Transistor Q_3 turns on, and by-passes a portion of Q_1 base current. Less current is available to Q_1 base, therefore TP 2 voltage increases. Transistor Q_1 has its conduction reduced, and the output voltage is decreased.
TP 2	=	23 V	
TP 3	=	10 V	
TP 4	=	6.5 V	
TP 5	=	12.5 V	
TP 6	=	11.8 V	

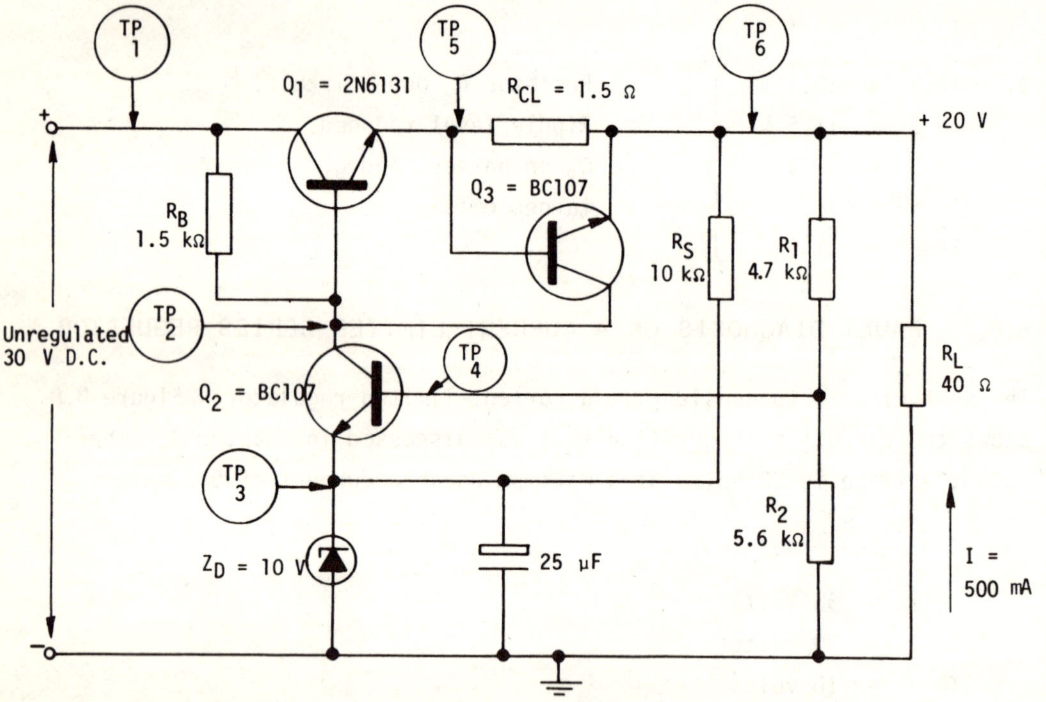

Figure 8.8 Current-limited series regulator

2.

TP 1	=	29 V	Excessive current is demanded (750 mA) and Q_3 base-emitter is open circuit. Ripple level is higher, D.C. output is only reduced slightly. Load could be damaged because circuit does not reduce the output voltage availability.
TP 2	=	20 V	
TP 3	=	10 V	
TP 4	=	10 V	
TP 5	=	19 V	
TP 6	=	19 V	

3. TP 1 = 32 V
   TP 2 = 1.2 V
   TP 3 = 10 V
   TP 4 = 750 mV
   TP 5 = 1.2 V
   TP 6 = 1.2 V

   Transistor Q_3 has a shorted base-emitter. Ripple level high. TP 2 pulled down and Q_1 turned off. Input voltage increased, indicating very little current being drawn.

4. TP 1 = 32 V
   TP 2 = 2.4 V
   TP 3 = 2 V
   TP 4 = 600 mV
   TP 5 = 2 V
   TP 6 = 1.2 V

   Current-sensing resistor R_{CL} open circuit. Input voltage increased, no current drawn. No output circuit, therefore no output voltage.

The next three fault conditions are for you to diagnose. The answers can be found in Appendix A at the back of the book.

5. TP 1 = 30 V
   TP 2 = 21.5 V
   TP 3 = 20 V
   TP 4 = 11.4 V
   TP 5 = 21 V
   TP 6 = 20.5 V

   Higher than normal ripple.
   Fault: ?

6. TP 1 = 33 V
   TP 2 = 33 V
   TP 3 = 0 V
   TP 4 = 0 V
   TP 5 = 0 V
   TP 6 = 0 V

   Fault: ?

7. TP 1 = 30 V
   TP 2 = 21.5 V
   TP 3 = 10 V
   TP 4 = 11.4 V
   TP 5 = 21.5 V
   TP 6 = 21 V

   Higher ripple level.
   Fault: ?

8.6.3 FAULT DIAGNOSIS ON A THREE-TERMINAL REGULATOR

A slightly different technique is adopted when diagnosing faults in three-terminal regulator circuits. The actual components of the device are not accessible for testing, and therefore test points are limited to the input, output, and third terminals. Figure 8.9 shows an adjustable three-terminal regulator circuit with test points. The output is adjusted to 25 volts, and therefore the load current is 500 mA.

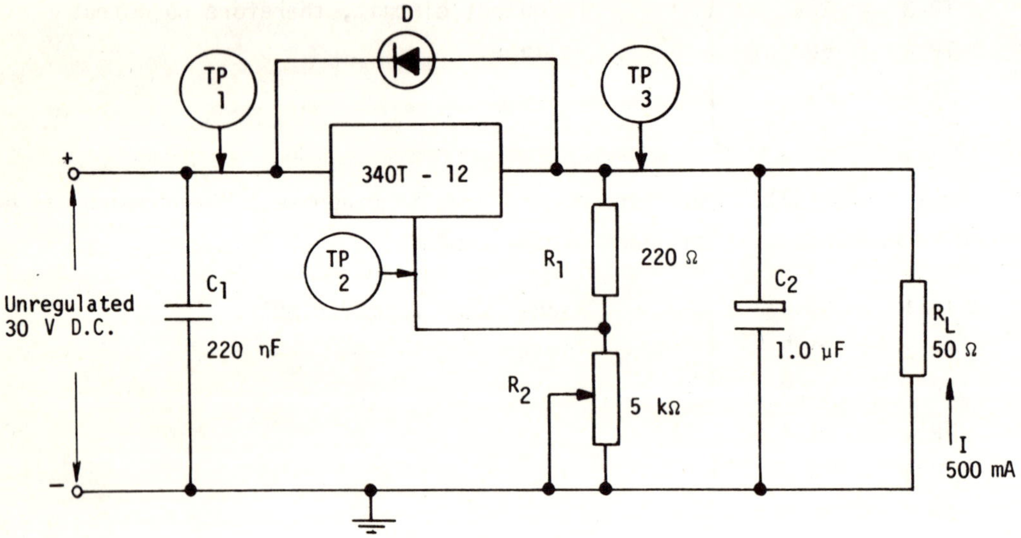

Figure 8.9 Three-terminal adjustable regulator

Normal working conditions

TP 1 = 30 volts
TP 2 = 13 volts
TP 3 = 25 volts
Ripple voltage, 10 mV p-p
Output current, 500 mA

Test point voltages under fault conditions

1. Diode open circuit: no change in circuit operation, but no input or output circuit protection.

2. Capacitor C_1 open circuit: may increase the ripple level, but no change in D.C. working conditions.

3. Capacitor C_2 open circuit: transient response poor, but no change in working conditions.

4.	TP 1 = 0 V TP 2 = 0 V TP 3 = 0 V	Fuse blown upon inspection, capacitor C_1 short circuit.
5.	TP 1 = 28 V TP 2 = 0 V TP 3 = 0 V	No D.C. output or adjustment voltage. Capacitor C_2 short circuit, or I.C. open circuit.
6.	TP 1 = 31 V TP 2 = 1.2 V TP 3 = 13 V	Resistor R_1 open circuit, Input D.C. increased, less current being drawn. No adjustment voltage fed back.
7.	TP 1 = 30 V TP 2 = 26.5 V TP 3 = 26.5 V	Resistor R_2 open circuit. Full adjustment voltage fed back. Output voltage and current increased. Ripple high.
8.	TP 1 = 29 V TP 2 = 17 V TP 3 = 29 V	Input and output voltages equal. Ripple high. Diode or I.C. short circuit.

With regulator circuits that employ feedback, i.e. 'closed loop' type, it is important to realize that a fault in the comparator or error amplifier can give a false fault symptom.

In a complex circuit it may be necessary to disconnect the feedback loop in order to perform checks on the basic regulating circuit test points.

8.6.4 FAULT DIAGNOSIS ON A SIMILATED SWITCHED-MODE POWER SUPPLY

The final circuit configuration to consider is the switched-mode circuit. Figure 8.10 shows a basic S.M.P.S. concept, using the variable resistance feature of a transistor to 'pulse-width modulate' a NE 555 timer I.C. The output of the NE 555 drives the base of the switching transistor.

Parts List.

Q_1	=	2N6131	$D_1 + D_2$	=	Power diodes
Q_2	=	BC107	L	=	41 mH (approx.)
R_1	=	2.2 kΩ	C_1	=	2.2 nF
R_2	=	10 kΩ pot.	C_2	=	1000 μF
R_3	=	10 kΩ	C_3	=	10 nF
R_4	=	68 kΩ			
R_5	=	100 Ω			
$R_6 + R_7$	=	1 kΩ			

The NE 555 can be pulse-width modulated by varying the voltage applied to Pin 5. As the voltage at Pin 5 is increased, the width of the mark increases.

Resistors R_1 and R_2 provide an adjustable 'feedback voltage' which is proportional to the output voltage. R_2 is adjusted to provide sufficient forward bias for transistor Q_2 to set the voltage level at Pin 5 of the I.C., and so drive transistor Q_1. As the value of the load resistor R_L is decreased, there is more demand on the power supply for current. The output voltage level falls, and hence the feedback voltage at Q_2 base is reduced. The forward bias (V_{BE}) of Q_2 is reduced, less collector current is drawn, and the potential at Q_2 collector rises. This increase in voltage is applied to Pin 5 of the I.C. The mark width increases, thereby increasing the 'on' time of transistor Q_1. Transistor Q_1 conducts for a slightly longer period of time, and this restores the output voltage to its former value.

Normal working conditions

TP 1 = 12 volts
TP 2 = 5.2 volts
TP 3 = 5.0 volts
TP 4 = 660 mV

TP 5 = 8.2 volts
TP 6 =
TP 7 =
TP 8 =

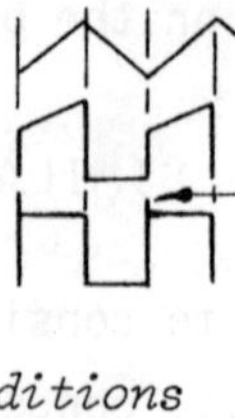

Inductive spike

These indicate the oscillator is working.

Output Voltage = 5 volts

Test point voltages under fault conditions

1. TP 1 = 10 V, TP 2 = 0, TP 3 = 0, TP 4 = 0; TP 5 = 9.5 V; TP 6 =, TP 7 =, TP 8 = } No Osc. } Multivibrator (NE 555) failed. No drive to switching-transistor base

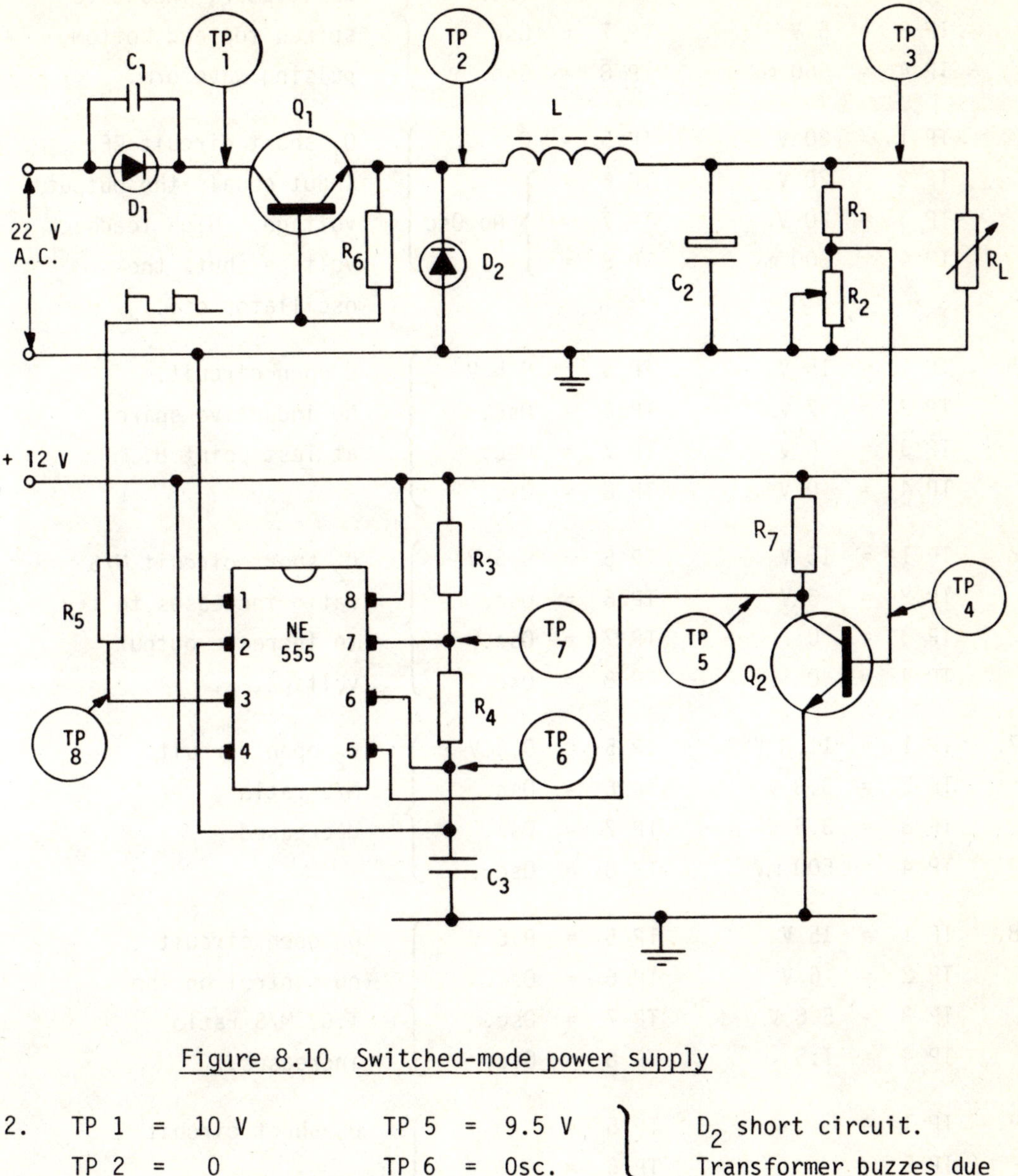

Figure 8.10 Switched-mode power supply

2.	TP 1	=	10 V	TP 5	=	9.5 V	}	D_2 short circuit.
	TP 2	=	0	TP 6	=	Osc.		Transformer buzzes due
	TP 3	=	0	TP 7	=	Osc.		to excess I. M/S ratio
	TP 4	=	0	TP 8	=	Osc.		increases.

3.	TP 1	= 12.5 V	TP 5	= 8.2 V	D_2 open circuit, unstable oscillator, inductive spikes top and bottom, pulsing waveform.
	TP 2	= 5.2 V	TP 6	= Osc.	
	TP 3	= 5 V	TP 7	= Osc.	
	TP 4	= 660 mV	TP 8	= Osc.	
4.	TP 1	= 20 V	TP 5	= 0	Q_1 short circuit CE. Input equals the output voltage. High feedback voltage shuts the oscillator off.
	TP 2	= 20 V	TP 6	= No Osc.	
	TP 3	= 20 V	TP 7	= No Osc.	
	TP 4	= 800 mV	TP 8	= No Osc.	
5.	TP 1	= 15 V	TP 5	= 9.5 V	L open circuit. No inductive spike at Test point 8.
	TP 2	= 7 V	TP 6	= Osc.	
	TP 3	= 0 V	TP 7	= Osc.	
	TP 4	= 0 V	TP 8	= Osc.	
6.	TP 1	= 10 V	TP 5	= 9.5 V	C_2 short circuit M/S ratio increases to try to increase output voltage.
	TP 2	= 3 V	TP 6	= Osc.	
	TP 3	= 0	TP 7	= Osc.	
	TP 4	= 0	TP 8	= Osc.	
7.	TP 1	= 11.8 V	TP 5	= 9.5 V	C_2 open circuit M/S ratio decreased.
	TP 2	= 3.8 V	TP 6	= Osc.	
	TP 3	= 3.6 V	TP 7	= Osc.	
	TP 4	= 500 mV	TP 8	= Osc.	
8.	TP 1	= 15 V	TP 5	= 9.5 V	Q_2 open circuit no control on the I.C. M/S ratio increases.
	TP 2	= 6 V	TP 6	= Osc.	
	TP 3	= 5.8 V	TP 7	= Osc.	
	TP 4	= 1.5 V	TP 8	= Osc.	
9.	TP 1	=	TP 5	=	Q_2 short circuit
	TP 2	=	TP 6	=	
	TP 3	=	TP 7	=	
	TP 4	=	TP 8	=	

10. TP 1 =	TP 5 =		R_5 open circuit.
TP 2 =	TP 6 =		No inductive spike
TP 3 =	TP 7 =		on oscillator waveform.
TP 4 =	TP 8 =		

The test point voltages in Faults 9 and 10 have been left blank, so that you can diagnose the expected voltages from the faults indicated. The answers can be found in Appendix A at the back of the book.

8.6.5 EXPERIMENTAL SWITCHED-MODE CIRCUIT

An S.M.P.S. supplied from a full-wave rectifier circuit and an I.C. comparator is shown in Figure 8.11. The output voltage is 'sampled' and compared to a reference voltage. The comparator output then controls the

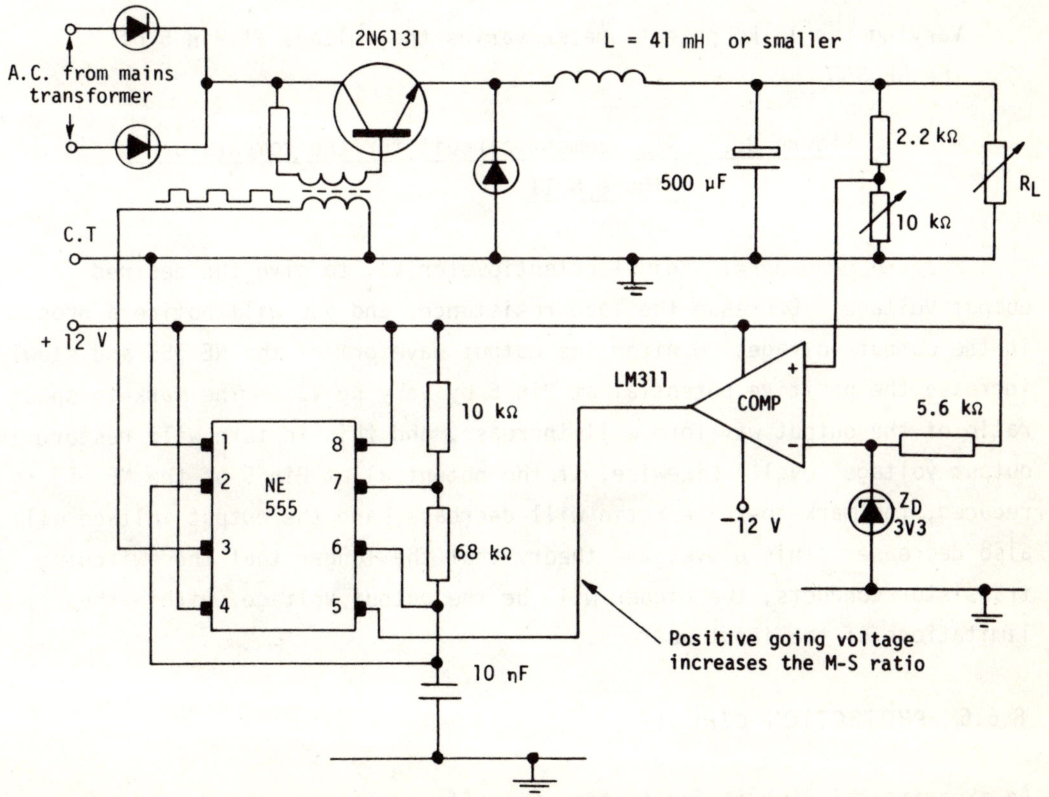

Figure 8.11 Experimental S.M.P.S. circuit

multivibrator circuit (NE 555). This circuit is easy to build and is ideal for the experimenter. If you want to observe the circuit action, the I.C. can be replaced with a potentiometer connected as shown in Figure 8.12.

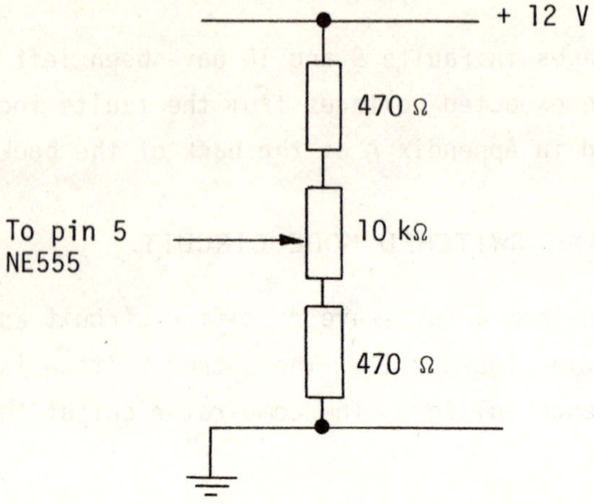

Varying the 10 kΩ potentiometer varies the voltage at Pin 5 of the NE 555

Figure 8.12 Replacement circuit for the comparator in Figure 8.11

Testing procedure. Adjust potentiometer V_{R1} to give the desired output voltage. Decrease the load resistance, and you will notice a drop in the output voltage. Monitor the output waveform of the NE 555 and slowly increase the positive potential at Pin 5 by varying V_{R1}. The mark-to-space ratio of the output waveform will increase, and this in turn will restore the output voltage level. Likewise, if the potential at Pin 5 of the NE 555 is reduced, the mark-to-space ratio will decrease, and the output voltage will also decrease. This proves the theory that the longer that the switching transistor conducts, the higher will be the output voltage, within the limitations of the input voltage.

8.6.6 PROTECTION CIRCUITS

An experimental circuit for testing the effect of over-voltage and excess current protection is shown in Figure 8.13. Transistor Q_1 is the switching transistor, and Q_2 is the driver, which in turn is switched on and off by a

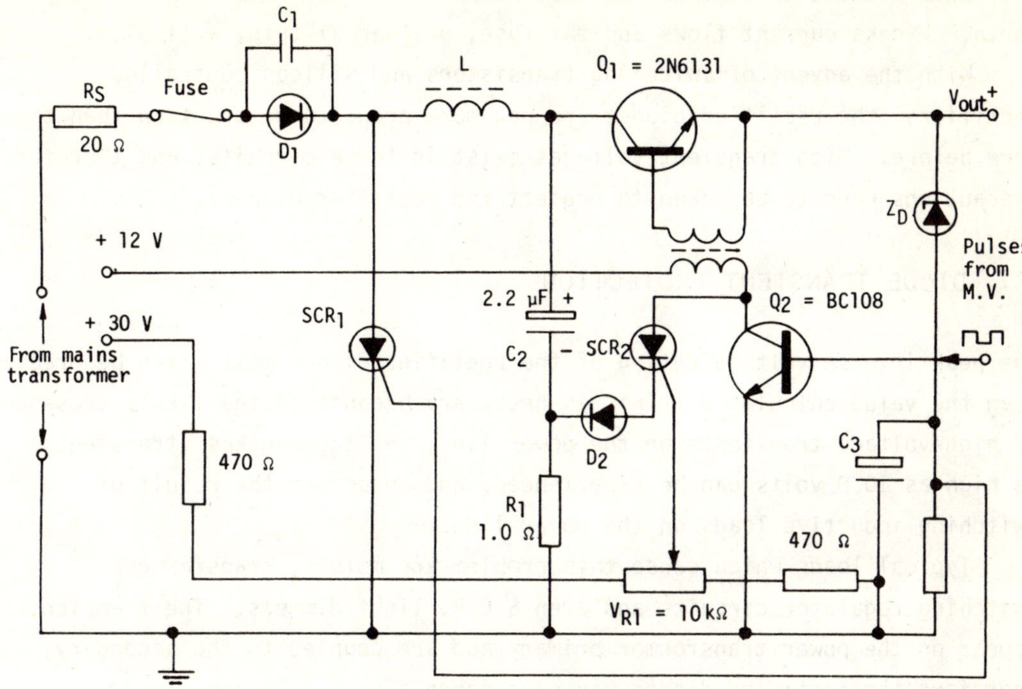

Figure 8.13 Experimental voltage and current-protection circuit

multivibrator circuit (M.V.) A silicon controlled rectifier (S.C.R.$_1$) is the crowbar protection to switch the input 'off' in the event of excess output voltage; this was discussed in Chapter 7. S.C.R.$_2$ provides protection in the case of excessive current.

Pulses from Q_1 collector are fed to S.C.R.$_2$ cathode via diode D_2. The gate voltage is set by V_{R1} to a potential just under the firing level. If a fault condition exists, excessive current will flow through Q_1. The negative-going current-pulses will drive the cathode of S.C.R.$_2$ negative with respect to the pre-set gate voltage. The S.C.R. will fire, and short out the supply to the collector of Q_2, thereby removing the drive to the switching transistor.

The over-voltage crowbar protection is connected directly across the supply voltage to the switching transistor. Normally the S.C.R. will not conduct, but if the output voltage exceeds the zener diode voltage, the zener will fire, capacitor C_3 will charge, and S.C.R.$_1$ will be switched on. This puts a short circuit on the input supply from the cathode on D_1 to earth. Excess current flows and the fuse, or thermal trip, will blow.

With the advent of switching transistors and silicon controlled rectifiers, the rectifier diodes are now more prone to destruction than they were before. High transient voltages exist in these circuits, and therefore, precautions need to be taken to protect the rectifier diodes.

8.7 DIODE TRANSIENT PROTECTION

The peak inverse voltage rating of the rectifier diodes must often be greater than the value calculated. This is necessary because of the likely presence of high-voltage transients on the power line. Voltage spikes (transients) as high as 1000 volts can be experienced, and these are the result of switching inductive loads on the power line.

Typical loads which cause this problem are motors, transformers, switching regulator circuits and even S.C.R. light dimmers. The transients appear on the power transformer primary and are coupled to the secondary, therefore the rectifier diodes may experience a very high peak voltage.

The simplest way to protect the diodes against these transients is to use high-current diodes, but these can be expensive. However, alternative methods are available, and these involve placing some line-impedance in the transformer primary or secondary circuit. This is slightly detrimental to the circuit's basic regulation, but it can be overcome.

Transient protection relies on shunting the transient around the rectifier diodes to dissipate the transient energy in the protection circuit.

Figure 8.14 shows a basic half-wave rectifier circuit complete with capacitive filter C_F. Resistor R, inductor L, and capacitors C_1 and C_2 all afford some protection to the diode. The combinations of components to provide transient protection are:

(a) Series resistor R and shunt capacitor C_1 in the primary circuit.

(b) Series inductor L and shunt capacitor C_1 in the primary circuit.

(c) Shunt capacitor C_2 across the secondary winding.

(d) Further protection can be obtained by shunting the diode with a capacitor (see Figure 8.15).

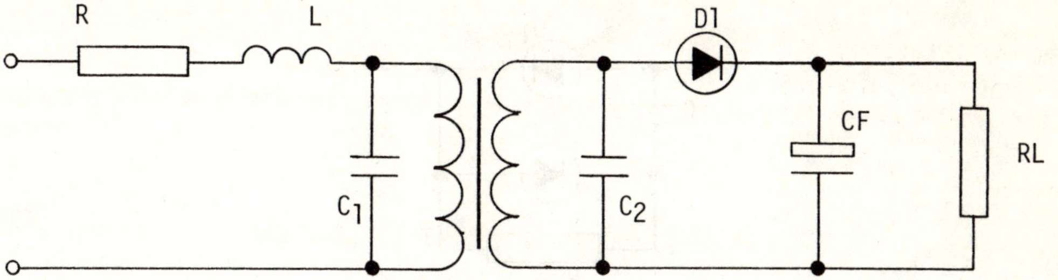

Figure 8.14 Basic half wave circuit with transient protection components

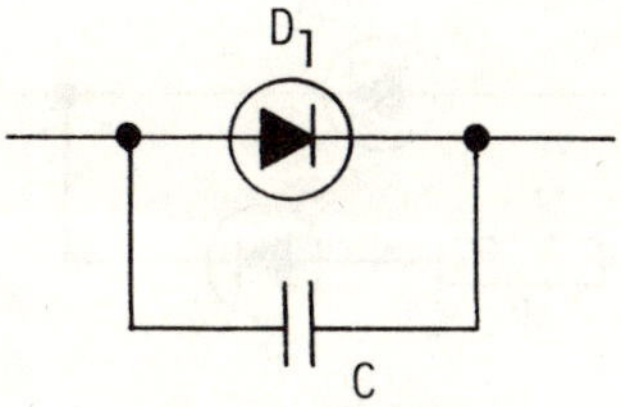

Figure 8.15 Rectifier diode shunted by C

(e) A surge suppression varactor can be placed across the diode (see Figure 8.16). This is very effective, but can be costly.

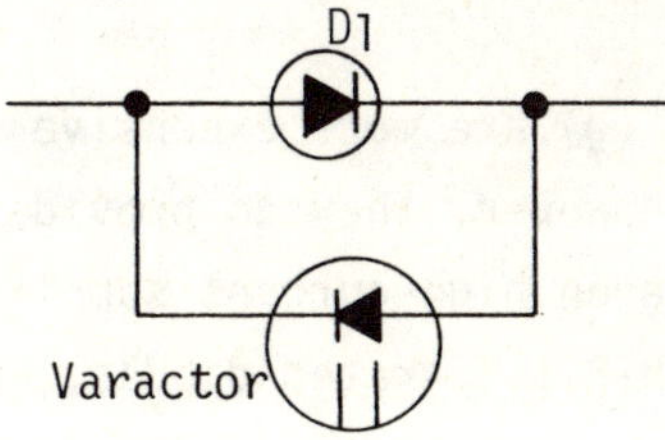

Figure 8.16 Varactor protection

(f) A shunt-connected dynamic-clipper circuit, consisting of a resistor, capacitor, and another diode. (see Figure 8.17)

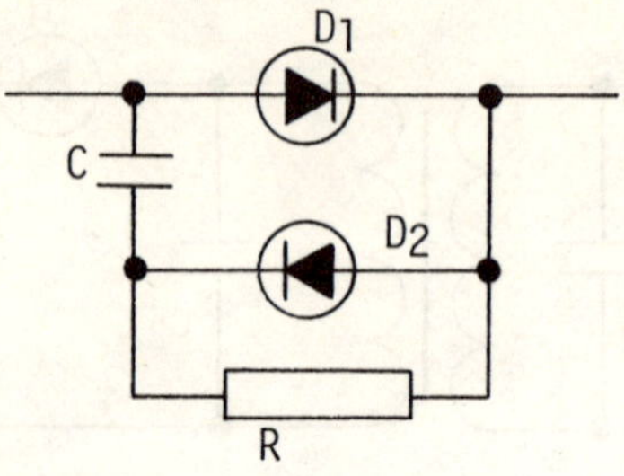

Figure 8.17 A dynamic clipper

(g) A zener-diode shunt across the rectifier diode; a series resistance may be included in the zener branch. (see Figure 8.18)

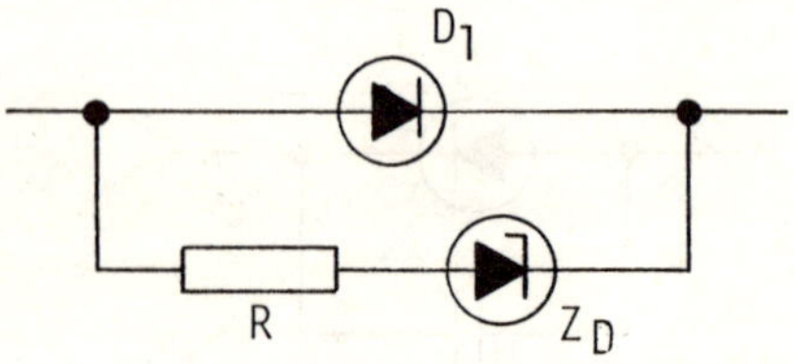

Figure 8.18 Zener diode shunt

8.7.1 COMPARISON OF PROTECTION CIRCUITS

Circuits (a), (b), (c) and (d) above are the cheapest, but they are limited in their ability to provide complete protection. Circuit (d) is the one that provides the best protection for the money, and it is quite suitable for low-current supplies.

Circuits (e), (f) and (g) are more expensive circuits because they use a further active device. However, they do provide the greatest protection, and their use is worthwhile on high-current supplies. They eliminated the need for expensive high-P.I.V.-rated diodes, and they successfully combat high transient voltages.

8.8 SELF-EVALUATION QUESTIONS (ANSWERS IN APPENDIX A)

In the text, we have looked at several circuit configurations, their operation, and protection circuits. For the final evaluation exercise, a new circuit configuration that has not been previously discussed is shown in Figure 8.19. It is a current-limited, adjustable-output, voltage regulator using a slightly different technique for error-signal amplification. Following is a list of parts for Figure 8.19:

D_1 - D_4		Power diodes			
Q_1 - Q_4		BC 107 or BC 108			
Q_5 -		2N6131 or 2N3055			
D_5 -		6.2 volt zener diode			
C_1	=	6400 μF 50 V.W.			
C_2	=	25 μF 10 V.W.			
C_3	=	470 nF			
R_1	=	1.5 kΩ	R_5	=	68 Ω
R_2	=	2.2 kΩ	R_6	=	1 Ω
R_3	=	1.0 kΩ	R_7	=	1 kΩ
R_4	=	470 Ω	R_8	=	1 kΩ
V_{R1}	=	1 kΩ	V_{R2}	=	100 Ω

Answer the following questions with reference to Figure 8.19.

1. Briefly explain the purpose of *all* the components listed in the parts list.

2. Explain the operation of the circuit if the load resistance decreases.

3. A fault develops in the power supply and the following test-point voltages result:

TP 1 = 23.8 V		TP 4 = 23.6 V
TP 2 = 23.6 V		TP 5 = 20.5 V
TP 3 = 11.9 V		

   What is the fault? Give reasons for your answer.

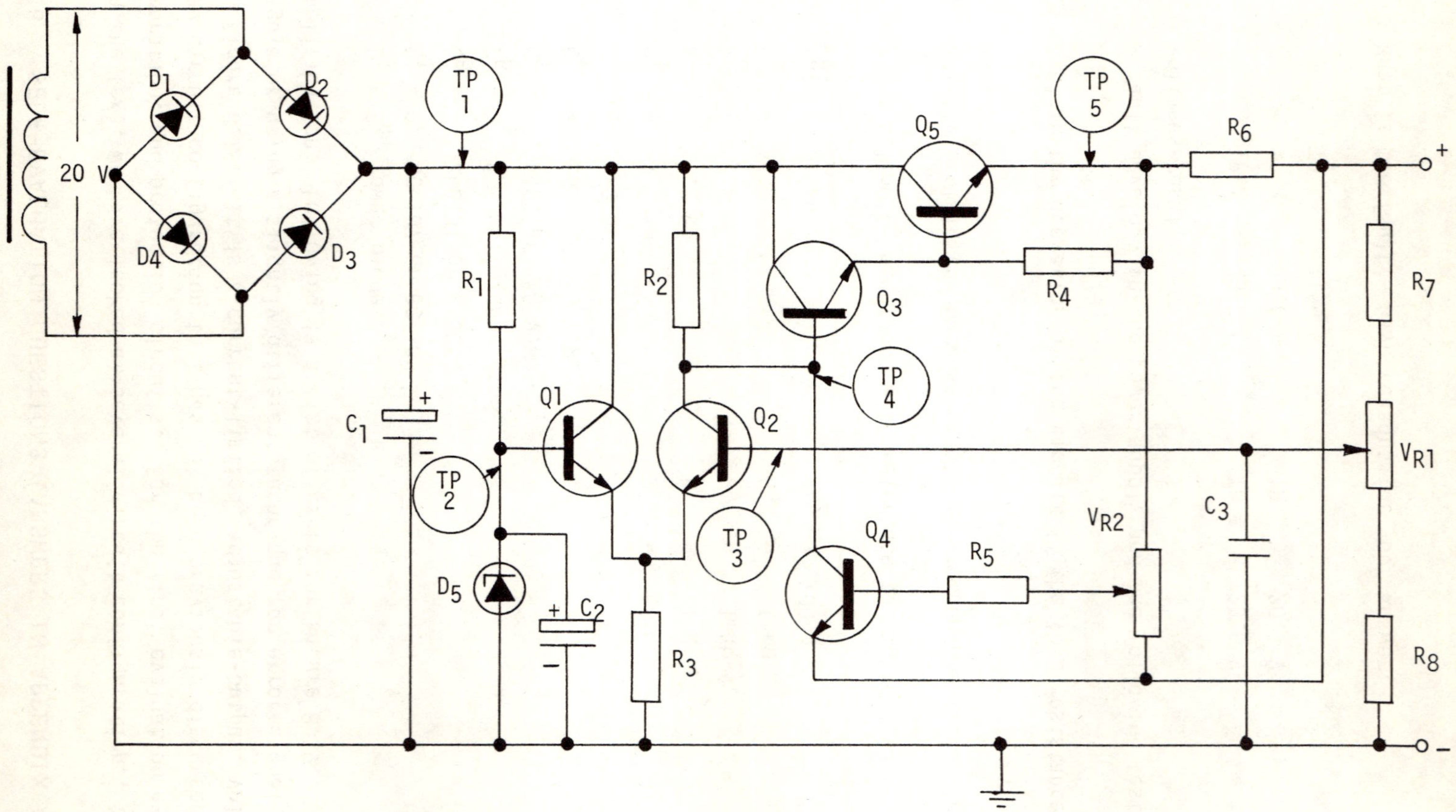

T.P.1 = 25 volts T.P.2 = 6.2 volts T.P.3 = 6.2 volts T.P.4 = 17.4 volts T.P.5 = 13.8 volts

Figure 8.19 Current-limited, adjustable output voltage regulator

4. Diagnose the fault given the following test-point voltages:

TP 1 = 25.6 V		TP 4 = 0	
TP 2 = 6.2 V		TP 5 = 0	
TP 3 = 0			

Give reasons for your diagnosis.

5. The power supply suddenly fails, and the following test-point voltage readings are taken:

TP 1 = 25.6 V	TP 4 = 25.6 V
TP 2 = 6.2 V	TP 5 = 0
TP 3 = 0	

What is the fault? Give your reasons.

6. Diagnose the fault from the following information, and support your diagnosis.

TP 1 = 24.5 V	TP 4 = 16 V
TP 2 = 0	TP 5 = 10.4 V
TP 3 = 4 V	

Appendix A

Answers to self-evaluation questions

CHAPTER 1

1. By reversing the diode

2. (i) Half

   (ii) All of

3. 700 millivolts

4. (a) V_S = 100 volts

   (b) P.I.V. = 142 volts

   (c) I_{Lm} = 14.14 amperes and $I_{L(RMS)}$ = 7.07 amperes

   (d) Power = 500 watts

5. 0.318 × 50 = 15.9 volts

6. (a) negative

   (b) (i) $V_{S1} + V_{S2}$ = 148.49 volts

   (ii) V_L = 94.5 volts

   (iii) I_L = 4.72 amperes

   (iv) 296.98 volts

7. (a) Full-wave bridge circuit.

   (b) D_2 and D_4 conduct. Point D is positive with respect to Point C.

   (c) (i) 36 volts

   (ii) 600 milliamperes

   (iii) 56.6 volts P.I.V.

8. (a) load voltage = 353.5 volts
   P.I.V. of diodes = 707 volts

   (b) Load voltage = 2 × 353.5 because the full secondary voltage is twice that of Part (a).

9. (a) Open circuit
   Short circuit
   High forward resistance

   (b) Open-circuit winding in either primary or secondary; shorted turn in either winding.

CHAPTER 2

1. To reduce the amount of A.C. ripple present in the load.

2. 0.29 per cent

3. (a) The average level increases

   (b) The diode current is reduced

4. Peak inverse voltage

5. 6 per cent

6. L-C Section

7. (a) $X_L = 3140\ \Omega$

   (b) $I = 3.8$ milliampere

   (c) Ripple voltage = 86.26 millivolt (peak)

8. (a) Use L-C filter

   (b) Use full-wave rectification

   (c) Use solid-state diodes

9. (a) 200 Hz

   (b) 400 Hz

CHAPTER 3

1. To maintain a constant output voltage with varying current demand on the circuit.

2. 66.6 Ω

3. 1.89 watts

4. Power wastage is high

5. 9.5 volts

6. (a) 72 per cent

   (b) 1.25 watts

   (c) 3.1 watts

7. (a) $R_L \downarrow I_L \uparrow V_E \downarrow V_{BE} \uparrow R_{CE} \downarrow V_{CE} \downarrow$ and V_E goes back to normal.

   (b) $V_{in} \uparrow V_E \uparrow V_{BE} \downarrow R_{CE} \uparrow V_{CE} \uparrow$ and V_E goes back to normal.

8. By shunting the base-emitter junction with a circuit that senses the load current and reduces V_{BE} when required.

9. Closed-loop circuit has an error-sensing amplifier and responds to very small changes in load variation.

10. To provide constant current with a varying load.
25 per cent
No

11. $I_L \uparrow I_E \uparrow V_{RS} \uparrow V_{BE} \downarrow \therefore I_L$ reduced to normal value.

CHAPTER 4

1. To decouple any noise on the input line.

2. (a) No

   (b) Improves transient response and noise rejection

3. (i) Mount for best convective cooling.

   (ii) Anodize or paint the heatsink surface.

   (iii) Use a thick heatsink material.

   (iv) Use a heat conductive paste on the I.C. pack.

   (v) Insulate the I.C. from the heatsink if necessary.

4. (i) $V_{in} - V_{out}$ differential - must not be too high or excessive power dissipation by the regulator itself.

   (ii) 'Drop-out' voltage - must not be too low or I.C. will lose its ability to regulate.

   (iii) Amplitude of the ripple - must not be excessive or it will intrude into the regulator drop-out voltage region.

5. The output current capability decreases, because the current-limit transistor requires a lower V_{BE} to turn it on at higher temperature levels.

6. A back-biased diode shunted across the regulator input/output. Yes, if V_{out} happened to exceed V_{in}, then the internal circuit of the I.C. is protected. Current path is via the diode.

7. Refer to Figure 4.9

8. By raising the ground pin above earth potential using a voltage divider from the output pin (See Figure 4.16).

9. No. Input and output capacitors require reverse polarity, and the short-circuit input protection diode must be reversed.

10. See Figure 4.30

11. A tracking regulator makes automatic adjustment for circuit variations. The positive regulator will follow or track the negative regulator or vice versa.
    Advantage: automatically keeps the positive/negative potentials within 100 mV of each other.

12. 1.5 V to 2.0 V.

13. (a) Power $= (V_{in} - V_{out}) \times I_L$
    $= 40$ watts.

    This will cause excessive heat to be dissipated by the regulator, thereby limiting its output current.

    (b) (i) Good heatsinking and extra convection cooling (i.e. a fan)

    (ii) Reduce V_{in} to lower the V_{in}-V_{out} differential.

CHAPTER 5

1. To step up D.C. voltages cheaply without the use of a step-up transformer.

2. Refer to Section 5.3

3. (a) $\sqrt{2} \times 2V_s$

   (b) $\sqrt{2} \times 2V_s$ for 3, $\sqrt{2} \times V_s$ for 1

   (c) Same as the input voltage frequency

4. Refer to Figure 5.12, and add one more section with diode polarity the same as D_2 and D_4 and with the capacitor C_6 on the bottom.

5. It is only suitable for small load currents.

6. T.V. picture tube (High V, low I)

CHAPTER 6

1. To provide a high D.C. output voltage from a low input D.C. voltage.

2. Not efficient at low voltages

3. An electronically controlled switch

4. Refer to Figure 6.5 and Section 6.3.2

5. By winding the primary and feedback windings 'bifilar'.

6. $$\text{Efficiency} = \frac{44}{50} \times 100 \text{ per cent}$$

   $$= 88 \text{ per cent}$$

   Yes, quite realistic.

7. By including a diode in the input lead which only conducts when the input polarity is correct.

8. Efficiency increases as the load increases.

CHAPTER 7

1. (i) High-frequency operation
   (ii) High-voltage o/p capacitor
   (iii) Switching regulator

2. (i) Efficient.
   (ii) Output voltage may be easily stepped up or down

3. (i) Current transients thrown back into the supply; heavy suppression required.
   (ii) Large output ripple level; can be reduced with adequate filtering.

4. The primary components could be at A.C. mains potential - i.e. not isolated.

5. Low equivalent series resistance (E.S.R.)

6. Refer to Figure 7.3 and Section 7.2.2

7. Step-down: transistor in series
   Step-up: transistor in parallel

8. Step-up circuit, because I flows in the load during both the 'on' and 'off' states of the transistor.

9. (a) Yes.
   (b) Inverting circuit with L replaced by a transformer.
   (c) Refer Figures 7.8 and 7.9.

10. Energy is stored in the inductor. The diode provides a current path through the load when the magnetic field around L collapses.

11. (i) A 'forward' S.M.P.S. transfers energy to the inductor and load when the transistor is on.

    (ii) A 'flyback' S.M.P.S. stores energy when the transistor is on and transfers it to the load when transistor is off.

    (iii) The 'forward' circuit has the highest power capability.

12. Refer to Section 7.5 and Figure 7.14

13. To prevent L from saturating due to the initial high current required to charge C_o.

14. It over rides the error amplifier output to keep the 'duty cycle' of the drive waveform to a minimum. This allows C_o to charge before the control circuit takes control of the circuit.

15. (i) Sensing resistor to provide a voltage proportional to output current, which then controls the error amplifier.

    (ii) Use the sensed voltage to inhibit the oscillator, and so remove the drive.

16. Refer Section 7.6.4 and Figure 7.26

17. The transistor is slow turning off. This dissipates power and may destroy the transistor.

18. Refer to Figure 7.29 and Section 7.6.6

CHAPTER 8

Faults in the current-limited series regulator

Fault 5: Open-circuit zener diode

Fault 6: Base emitter of Q_1 open circuit

Fault 7: Error-amplifier transistor Q_2 open circuit

Faults in the S.M.P.S. circuit

Fault 9:

TP 1 = 10 V	TP 5 = 0	
TP 2 = 0	TP 6 =	No osc.
TP 3 = 0	TP 7 =	No osc.
TP 4 = 0	TP 8 =	No osc.

Fault 10:

TP 1 = 10 V	TP 5 = 9.5 V	
TP 2 = 0	TP 6 = osc.	
TP 3 = 0	TP 7 = osc.	
TP 4 = 0	TP 8 = osc.	No inductive spike.

Self-evaluation questions

1. Purpose of the components in Figure 8.19, in order through the circuit.

- D_1 - D_4 - Full-wave bridge rectifier
- C_1 - Filter capacitor
- R_1 - Series-biasing resistor for the zener diode D_5
- D_5 - Reference diode to hold the base of Q_1 steady at 6.2V
- Q_1 & Q_2 - Differential pair error amplifier
- R_3 - Common emitter-resistor
- R_2 - Load resistor for error amplifier
- Q_3 - Control transistor which drives the series transistor Q_5. It in turn has its conduction determined by the output of the Q_1/Q_2 pair.
- Q_4 - Current-limiting device. When forward biased, it draws current away from Q_3 base.
- R_4 - Fixed leg of the voltage divider, R_4/Q_3, which controls the forward bias on Q_5.
- R_5 - Base current-limiting resistor for Q_4
- R_6 - Current-sensing resistor, across which the forward bias for Q_4 is developed.

R_7, R_8 - Fixed top and bottom legs of the voltage divider across the output. They prevent the output from being shorted to earth when the output voltage is adjusted.

V_{R1} - Adjustment for the output voltage. Varies the voltage fed to the base of Q_2 which is compared at the base of Q_1.

V_{R2} - Adjustment for setting the current limit. It is in parallel with the sensing resistor R_6.

C_3 - High frequency bypass capacitor.

Circuit Operation: The reference diode is connected to transistor Q_1 base and holds it at a steady 6.2 volts D.C. A portion of the output voltage, determined by the setting of V_{R1}, is applied to the base of Q_2. Any difference between these two voltage levels is amplified and then applied to the base of Q_3.

The conduction of transistor Q_3 is varied, and this adjusts the forward bias on Q_5, thereby controlling the output level. Transistors Q_3 and Q_5 are connected in darlington configuration, giving a high gain to the circuit, making it very sensitive. Current limiting is provided by R_6 in parallel with V_{R2}. The voltage developed across this resistor network adjusts the forward bias on Q_4. As Q_4 conducts harder, it draws current away from Q_3 base. This reduces the conduction of Q_3, and hence its resistance (R_{CE}) increases. The forward bias for Q_5 is developed across R_4 (the lower leg of the Q_3/R_4 voltage divider) and is also reduced, therefore the conduction of Q_5 is less, resulting in a reduced output voltage.

If the load is decreased, the output current demand is less, and therefore Q_4 does not conduct, resulting in no current-limiting.

2. If the load current increases sufficiently, the voltage developed across R_6/V_{R2} will increase above 600 mV, and forward bias Q_4. Base current will be drawn away from Q_3, and it will conduct less, increasing in resistance. The forward bias on Q_5 is therefore reduced; the resistance increases, resulting in more voltage being dropped across Q_5, and the output voltage is reduced.

3. Fault: zener diode open circuit. The voltage at TP 1 has dropped, indicating that more current is being drawn. All the other voltages

have increased. Voltage at TP 2 has increased, indicating that the zener diode is inoperative. Q_1 conducts heavily, Q_2 conduction ceases, and therefore the base voltage at Q_3 increases. Q_5 is turned on harder and the output voltage increases.

4. Fault: Q_5 open circuit. The voltage at TP 1 has increased, indicating that less current is being drawn. The reference voltage, derived from the input, is normal, and all other voltages are zero. Q_5 is non-conducting because it is open circuit, therefore no output voltage.

5. Fault: transistor Q_3 open circuit. The voltage at TP 1 has increased indicating that less current is being drawn. TP 2 is normal, but TP 4 is high, indicating no current through R_2. The output and feedback voltages are zero because Q_5 is not being driven by Q_3.

6. Fault: zener diode or C_2 short circuit. The voltage at TP 1 has dropped indicating that more current is being drawn. TP 2 is zero. The output of the differential pair (TP 4) is low, therefore Q_3 and hence Q_5 have their conduction reduced. The output voltage is low. Ohmmeter check from TP 2 to earth shows a short circuit.

Appendix B

Glossary of terms and electronic symbols

GLOSSARY OF NOTATION AND TERMS

Average	Represents the D.C. value of a waveform, and can be calculated as follows: 1. 0.637 times the peak value of a full wave. 2. 0.318 times the peak value of a half wave.
CL	Current limiting
E.S.R.	Equivalent series resistance
I_B	Base current
I_C	Collector current
I_E	Emitter current
I_L	Load current
I_Q	Quiescent current
I_Z	Zener diode current
Inverted	The output is the opposite polarity to the input
Peak	Maximum value
P.I.V.	Peak inverse voltage. The maximum voltage a diode can withstand in the reverse bias direction.
R.M.S.	Root mean square. The effective value of an A.C. waveform, and can be calculated as follows: 1. 0.707 times the peak value of a full wave. 2. 0.5 times the peak value of a half wave.
Ripple	Fluctuations above and below the average level
R_{CE}	Collector - emitter resistance of a transistor
Reg	Regulator

R_L	Load resistance
Step-down	The secondary voltage is less than the primary voltage.
Step-up	The secondary voltage is greater than the primary voltage.
SMPS	Switched mode power supply
Thermal Shutdown	Circuit ceases to operate because of an increase in temperature.
Time Constant (T.C.)	The time taken for the circuit voltage or current to change by 63 per cent of its former steady value. It can be calculated as follows: 1. T.C. = C × R (capacitive circuit). 2. T.C. = L/R (inductive circuit).
T_j	Junction temperature of an I.C
TP	Test point
V_B	Base voltage
V_C	Collector voltage
V_E	Emitter voltage
V_{BE}	Base - emitter voltage of a transistor
V_{CE}	Collector - emitter voltage of a transistor
V_{CB}	Collector - base voltage of a transistor
V_L	Load voltage
V_{NL}	Voltage output with no load
V_{FL}	Voltage output with maximum load
$V_{dropout}$	The difference between the input and output voltage at which the regulator ceases to function
V_{Reg}	The output voltage of the regulator
V_Z	Zener diode voltage
Z_D	Zener diode

TABLE OF ELECTRONIC SYMBOLS

Symbols used in the text for electronic components are shown below.

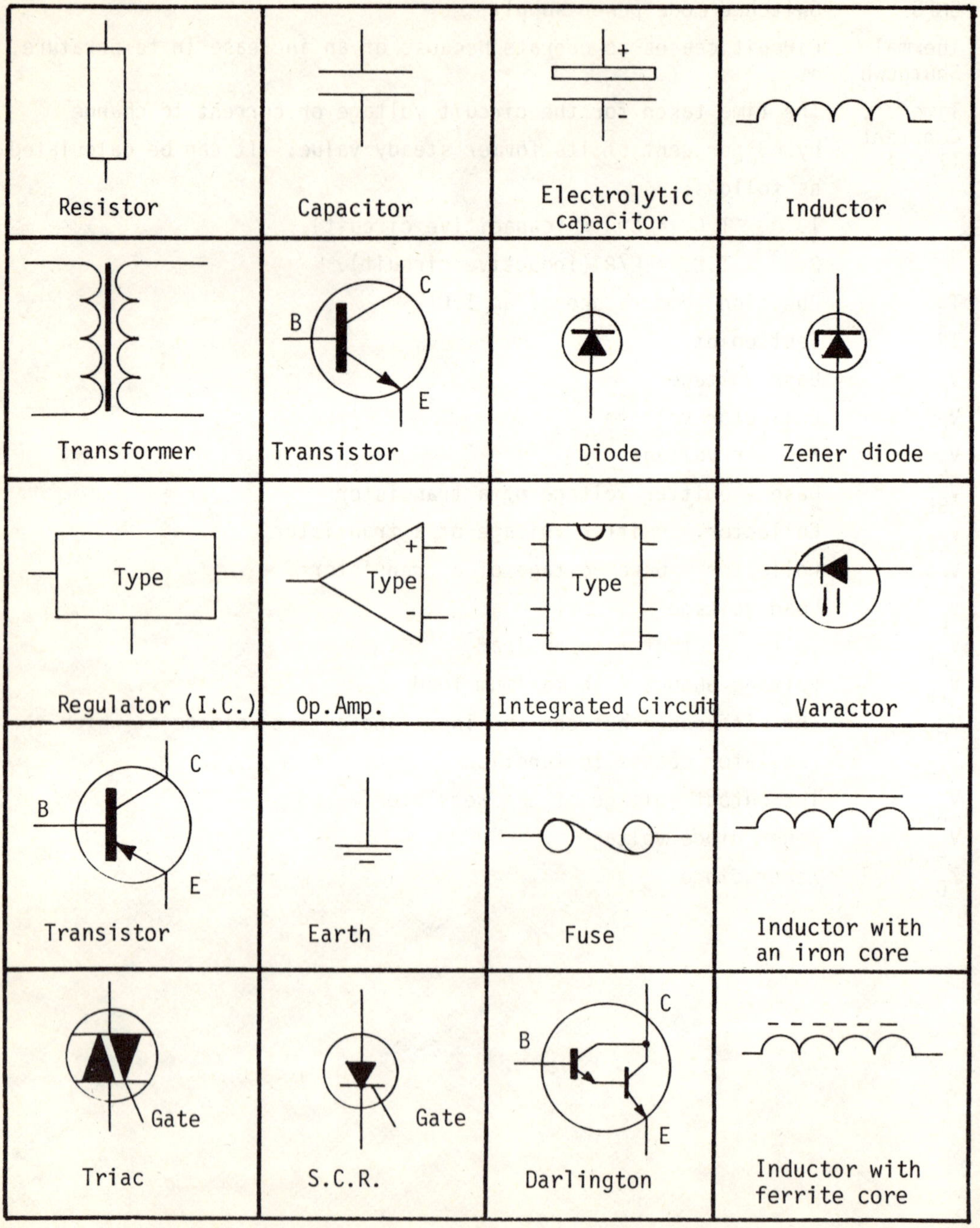

Appendix C

Calculation of ripple and form factors for unfiltered half and full wave rectifiers

The output of a power supply contains both D.C. and A.C. voltage components. During the design stage of a power supply, we must be aware of the relationship between the A.C. voltage component and the D.C. output voltage.

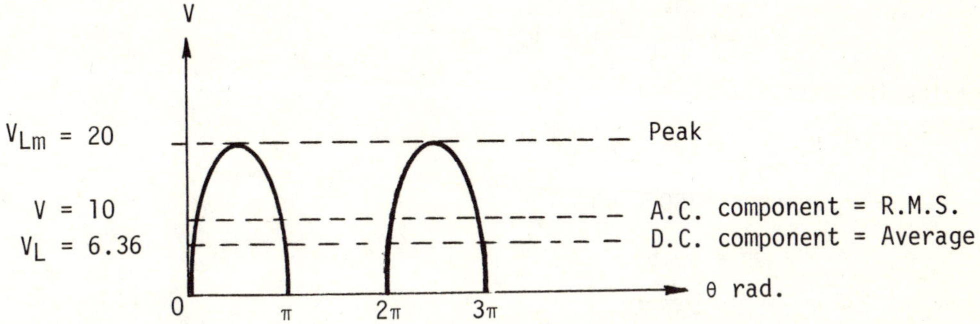

Figure C.1 Output of a half-wave rectifier circuit

The peak voltage of the waveform in Figure C.1 is 20 V, therefore:

D.C. component = 0.318 × 20

= 6.36 volts.

The R.M.S. value of the waveform

(complete load voltage) = *0.5 × 20

= 10.0 volts

* See Appendix D

This value is the R.M.S. value of the complete waveform, which contains both the A.C. and D.C. voltages. Complex mathematics can show that the R.M.S. value of the whole waveform is equal to the square root of the sum of the squares of each R.M.S. component in the waveform:

$$V = \sqrt{V_1^{\,2} + V_2^{\,2}}\ .$$

Therefore:

$$V_{Load} = \sqrt{V_{A.C.}^{\;2} + V_{D.C.}^{\;2}}$$

where $V_{A.C.}$ = R.M.S. value of A.C. component,

and $V_{D.C.}$ = R.M.S. value of D.C. component.

$V_{D.C.}$ = 6.36 volts, and since the R.M.S. value of a D.C. component is equal to the D.C. component itself, then:

$$V_{Load} = \sqrt{V_{A.C.}^{\;2} + 6.36^2}$$

$$10 = \sqrt{V_{A.C.}^{\;2} + 6.36^2}.$$

Solving the equation:

$$V_{A.C.}^{\;2} = 10^2 - 6.36^2$$

$$= 100 - 40.45$$

$$= 59.55$$

$$V_{A.C.} = \sqrt{59.55}$$

$$= 7.71 \text{ volts}$$

This 7.71 volts is the R.M.S. value of the ripple voltage. Therefore the load voltage of the unfiltered rectifier output contains a D.C. component = 6.36 volts, and an A.C. component = 7.71 volts (R.M.S.)

Ripple factor (R.F.). The ripple factor is an important circuit parameter; it is the ratio of ripple voltage (R.M.S.) to D.C. component.

$$R.F. = \frac{V_{A.C.}}{V_{L(D.C.)}}$$

For the waveform in Figure C.1:

$$R.F. = \frac{7.71}{6.36}$$

$$= 1.21 \text{ or } 121\%.$$

Form factor (F). A further measurement of power-supply performance is the ratio of R.M.S. load voltage to the D.C. component. This is called the form factor. For the waveform in Figure C.1, the R.M.S. load voltage (complete A.C. and D.C. components) is equal to 10 volts, and the D.C. component is equal to 6.36 volts.

Therefore: $F = \frac{V}{V_L}$ or $\frac{V_{L(A.C.+D.C.)}}{V_{L(D.C.)}}$

$= \frac{10}{6.36}$

$= 1.57$

In an ideal power supply, the A.C. component would be zero, resulting in a form factor of VL/VL or F = 1. However, practically this does not happen, and an efficient filtering circuit reduces the amplitude of the ripple in the D.C. output.

The same theory applies in determining R.F. and F for a full-wave rectifier circuit (see Figure C.2).

In order to reinforce the difference between a half- and a full-wave rectifier, the same peak voltage will be used to calculate R.F. and F.

D.C. component $= 0.637 \times 20$

$= 12.74$ volts

R.M.S. value of the waveform $= 0.707 \times 20$

$= 14.14$ volts

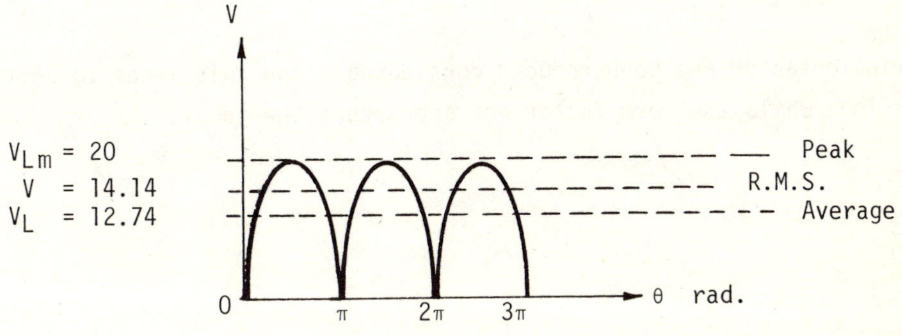

Figure C.2 Output of a full-wave rectifier circuit

Since: $$V_{Load} = \sqrt{V_{A.C.}^2 - V_{D.C.}^2}$$

Then:

$$14.14 = \sqrt{V_{A.C.}^2 - 12.74^2}$$

Solving:

$$(V_{A.C.})^2 = 14.14^2 - 12.74^2$$

$$= 199.94 - 162.3$$

$$= 37.63$$

$$V_{A.C.} = \sqrt{37.63}$$

$$= 6.13 \text{ Volts.}$$

The ripple factor is:

$$R.F. = \frac{V_{A.C.}}{V_{L(D.C.)}}$$

$$= \frac{6.13}{12.74}$$

$$= .48 \text{ or } 48\ \%.$$

The 'form factor' is:

$$F = \frac{V}{V_L}$$

$$= \frac{14.14}{12.74}$$

$$= 1.1.$$

The ripple factor has been reduced considerably and this leads to better filtering, while the form factor now approaches the ideal.

Appendix D

Calculation of average and R.M.S. values

AVERAGE VALUE

The average value of a repetitive waveform is also referred to as the D.C. component. It can be found by algebraically summing the area under the waveform, and dividing by the total time period:

$$F_{av} = \frac{1}{T} \int_{o}^{T} f\ (t)\ dt$$

F_{av} = average value of f (t)

Example 1. Find the average value of the waveforms in Figure D.1 (a) and (b).

(a) Total period for one repetition (T) is 2 seconds. Finding the values for individual time segments, we get:

from 0 to 1 = 1 × 2 V = 2 V

from 1 to 2 1 × -2 V = -2 V.

Therefore the D.C. component (average value) is found to be:

$$\text{D.C. component (average value)} = \frac{\text{total area}}{\text{period}}$$

$$= \frac{2 - 2}{2}$$

$$= 0.$$

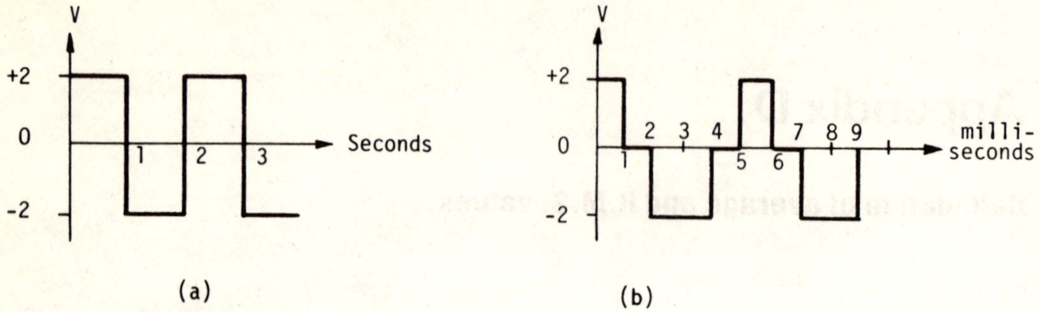

Figure D.1 Example waveforms

(b) Total period for one repetition is 5 milliseconds. The values for individual time segments are:

$$\text{from 0 to 1} = 1 \times 10^{-3} \times 2\ \text{V}$$
$$= 2 \times 10^{-3}\ \text{V}$$

$$\text{from 1 to 2} = 0$$

$$\text{from 2 to 4} = 2 \times 10^{-3} \times -2\ \text{V}$$
$$= -4 \times 10^{-3}\ \text{V}$$

$$\text{from 4 to 5} = 0$$

Therefore the D.C. component (average value) is:

$$\text{D.C. component (average value)} = \frac{\text{total area}}{\text{period}}$$

$$= \frac{(2 \times 10^{-3}) - (4 \times 10^{-3})\ \text{V}}{5 \times 10^{-3}}$$

$$= \frac{-2 \times 10^{-3}}{5 \times 10^{-3}}$$

$$= -0.4\ \text{V}$$

$$= -400\ \text{mV}$$

With power supplies, we are chiefly concerned with sinusoidal waveforms. A sinewave is symmetrical above and below the zero axis; therefore, the positive going region cancels the negative going region (as in Figure D.1a), and the D.C. component equals zero.

Our concern with the D.C. component is to determine the average voltage output level from a full-, or a half-wave rectifier. Figure D.2 shows the two output voltage waveforms.

Example 2. Find the average value of the waveforms in Figure D.2 (a) and (b).

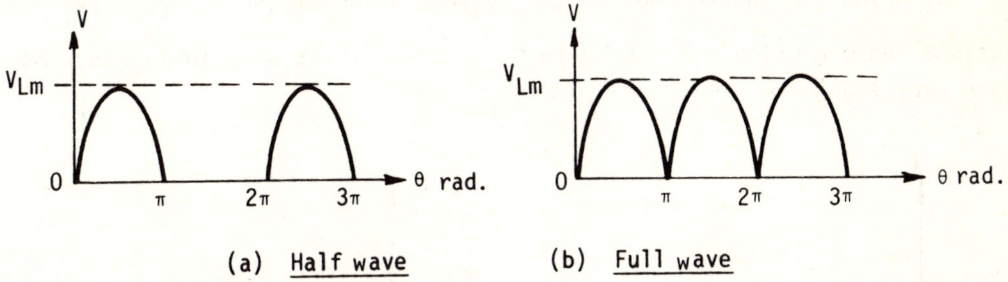

Figure D.2 Example waveforms

(a) Total period for one repetition is 2π radians. From 0 to π = $2V_{Lm}$ (from integration). From π to 2π = 0. Therefore the D.C. component (average value):

$$= \frac{2V_{Lm}}{2\pi}$$

$$= 0.318\ V_{Lm}$$

Hence, the average value of a half-wave rectified sinewave is equal to 0.318 of the maximum value.

(b) Total period for one repetition is 2π radians. From 0 to π = $2V_{Lm}$ (from integration). From π to 2π = $2V_{Lm}$. Therefore the D.C. component (average value):

$$= \frac{4V_{Lm}}{2\pi}$$

$$= 0.6366$$

$$= 0.637 \text{ (to 3 decimal places)}$$

Hence, the average value of a full-wave rectified sinewave is equal to 0.637 of the maximum value.

R.M.S. VALUE

The R.M.S. value of a repetitive waveform is also referred to as its effective or working value. It can be found in the following manner:

1. Square the waveform.
2. Find the average value of the squared waveform.
3. Find the square root of the average value of the squared waveform.

Parts a and b of Figure D.3 show the output waveforms of a full- and a half-wave rectifier.

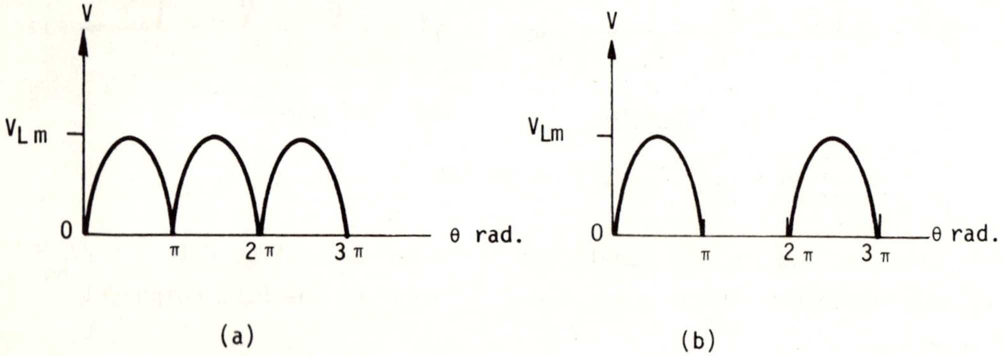

Figure D.3 Full- and half-wave output voltage waveforms

$$F = \sqrt{\frac{1}{T} \int_0^T f^2 (t)\, dt}$$

(F = R.M.S. value of f(t))

Equation for the waveform: $v = V_{max} \sin \theta$

Equation for the squared waveform: $v^2 = V_{max}{}^2 \sin^2\theta$

Solution. (a) The waveform in Figure D.3a is the output of a full-wave rectifier. The first step is to square this waveform, as shown in Figure D.4. It has the same R.M.S. value as a full sinewave. The waveform is symmetrical, therefore the average is half way between 0 and $V_{Lm}{}^2$, or $V_{Lm}{}^2/2$. Then we take the square root of the average value:

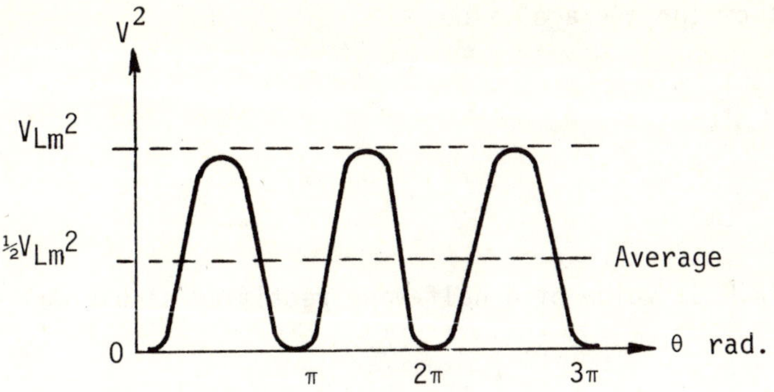

Figure D.4 Waveform in Figure D.3a squared

$$= \sqrt{\frac{V_{Lm}^2}{2}}$$

$$= \frac{V_{Lm}}{\sqrt{2}}$$

$$= 0.707 \times V_{max}$$

This is true for both a full sinusoidal waveform, and a full-wave rectified sinusoidal waveform.

(b) The waveform in Figure D.3b is the output of a half-wave rectifier. Again, the first step is to square this waveform. Due to the absence of

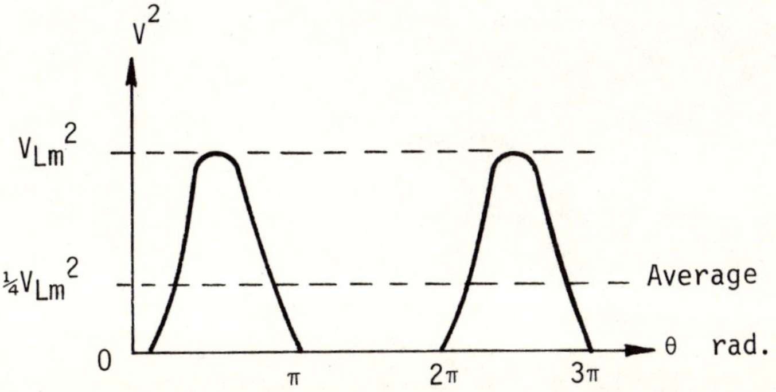

Figure D.5 Waveform in Figure D.3b squared

each second half cycle from the waveform, the average must be half the average of a squared full wave. Therefore the average value = $\frac{1}{4} V_{Lm}^2$. The square root of the average value:

$$= \sqrt{\tfrac{1}{4}V_{Lm}^2}$$

$$= \tfrac{1}{2}V_{Lm}.$$

Therefore, the R.M.S. value of a half-wave rectified sinusoidal waveform is $0.5 \times V_{max}$.

Bibliography

TEXTS AND MANUALS

R. Boylestead and L. Nashelsky, *Electronic Devices and Circuit Theory*, Prentice-Hall Inc., Englewood Cliffs, New Jersey, 1978.

H.H. Gerrish, *Transistor Electronics*, Goodheart-Willcox Co., Inc., Illinois, 1969.

B. Grob, *Basic Electronics*, McGraw-Hill Kogakusha Ltd., 1977.

R.G. Hibberd, *Transistor Pocket Book*, Newnes Books, Middlesex, U.K., 1968.

G.J. King, *Radio Circuits Explained*, Newnes Technical Books, London, 1977.

A.S. Manera, *Solid State Electronic Circuits for Engineering Technology*, McGraw-Hill Kogakusha Ltd., 1973.

N.S. Electronics, *Voltage Regulator Handbook 1978*.

D.J. Seal, *P.A.L. Receiver Servicing*, Foulsham Technical Books, Slough, Bucks., U.K., 1971.

M. Slurzberg and W. Osterheld, *Essentials of Communication Electronics*, McGraw-Hill Kogakusha Ltd., 1973.

Zarach and Morris, *Television Principles and Practice*, MacMillan, London, U.K., 1979.

JOURNAL ARTICLES

P. Bardos, 'Switched-mode P.S.V. design', *Electronics Industry* 5, No. 4, April, 1979, pp. 25-29.

M. Burchall, 'Why switching power supplies are rivalling linears', *Electronics* 51, No. 19, September, 1978, pp. 141-143.

B. Dance, 'I.C. of the month', *Practical Wireless* 55, No. 874, December, 1979, pp. 59-63.

R.C. Frostholm, 'One-chip controller simplifies switched-mode supplies', *Electronics* 52, No. 13, June, 1979, pp. 140-143.

Index

Abbreviations used in
  rectifier circuits, 2
  diode characteristics, 14
Adjustable power supplies
  transistor, 52
  I.C., 82, 89
Average value (calculation of), 220

Bridge rectifier, 9
  calculations, 13
  circuit, 11
  operation, 9
  waveforms, 11
Battery chargers, 93, 94

Capacitor filter, 23, 25
  circuit, 23, 26
  waveforms, 24, 27
  choice of capacitor, 28
Circuit measurments, 169
Component faults and replacement, 170
Control circuits for S.M.P.S., 145, 146, 147, 148
C-R filter, 33
  circuit, 33
  waveforms, 33
Crow-bar protection, 158
Current limiting, 54, 55, 56, 57, 155
Current regulators, 58,
Current source, 79, 84

D.C. to D.C. converters, 112
  applications, 121
  input polarity protection, 121
  push-pull transistor, 117
  single transistor, 115
Diode, 13
  calculations of P.I.V., 6, 9
  cathode identification, 14
  characteristics, 13
  protection, 196
  testing, 15
Doubler, 101
  circuit, 101, 103
  operation, 101, 103
Drop-out voltage, 71

Dual-polarity power supply, 83, 87
Dual complementary supply, 95

Efficiency
  D.C. to D.C. converter, 121
  linear regulator, 50
  S.M.P.S., 139, 141, 144
Electronic symbols, 215
Experimental circuits, 87, 93, 163, 164, 193, 195

Fault diagnosis
  integrated regulator, 188
  series regulator, 183, 185
  S.M.P.S., 189
Fault-finding procedure, 175
Fault symptom tables, 180, 182
Filtering, 21
  capacitor, 23, 25
  C-R section, 33
  L-C section, 30
  π section, 33
Fixed-output regulators
  I.C., 78, 83
  transistor, 49, 56, 57
Form factor (full- and half-wave), 216
Full-wave rectifier, 7
  calculations, 9
  circuit, 7
  operation, 8
  waveforms, 8, 9

Half-wave rectifier, 3
  calculations, 5
  circuit 3,
  operation, 3
  waveforms, 4, 5
Heatsinking, 68
High-current regulator, 80, 84

Integrated voltage regulators, 63
  application, 91
  connection diagrams, 66
  handling precautions, 76
  heatsinking, 68
  operation, 73
  packages, 64
  protection, 73, 77
  selection, 65
  use in S.M.P.S., 161

Junction temperature, 76
  effect on power capability, 76

L-C filter, 30
  circuit, 30
  waveforms, 31

Mains-input protection, 161

Negative I.C. regulator, 83
Operational amplifier as error amplifier, 56
Overload protection
  series transistor regulator, 54
  switched-mode regulator, 153
Over-voltage protection, 158

Packages (I.C.), 66
Peak inverse voltage, (P.I.V.), 2, 6, 9

Pi-filter, 33
  circuit, 33
  waveforms, 33
Positive I.C. regulator, 78
Power regulators, 89
Power-supply testing, 176
Printed circuit board repair, 174

Quadrupler, 108
  circuit, 109
  operation, 108

R.M.S. Value (calculation of), 220
Rectifiers, 1
  bridge, 9
  half-wave, 3
  full-wave, 7
Regulation
  current, 58, 178
  voltage, 36, 178
Ripple factor (full- and half-wave), 216
Ripple percentage, 22

Safety precautions, 171
Series transistor regulator, 48
  adjustable, 52
  efficiency, 50
  fixed, 49, 55
  operation, 50
Shunt transistor regulator, 47
  circuits, 47, 48
  operation, 47
Soft-start circuits (S.M.P.S.), 153
Surge-current protection, 157
Switched-mode supplies, 126
  basic operation, 131
  calculations, 140
  inverting, 135
  protection circuits, 153
  step-down, 131
  step-up, 134

Testing procedure
  capacitor, 37
  diode, 15
  inductor, 37
  transformer, 16
  voltage regulator circuits, 176, 181
Tracking regulator, 88
Transformer
  purpose, 1
  testing, 16
Transient suppression, 78
Tripler, 104
  circuit, 104, 106
  operation, 105, 107

Variable output regulators
  I.C., 82, 85, 89
  transistor, 52
Vibrator supplies, 113
Voltage multiplication, 99
  doubler, 101, 103
  tripler, 104
  quadrupler, 108
Voltage regulation, 36, 178
Voltage regulators
  I.C., 64

Voltage regulators *(continued)*
series, 48
shunt, 47
switched-mode, 126

Zener diode, 42
calculation of series resistor, 44
characteristics, 43
operation, 44
precautions, 46